Cheap AND *Easy!*

OVEN & COOKTOP REPAIR

2000 EDITION

Written ESPECIALLY for Trade Schools, Do-It-Yourselfers, and other "green" technicians!

By Douglas Emley

EB Publishing, Inc.

Carson City, Nevada ● Phone / Fax toll free 888-974-1224 ● website: http://www.appliancerepair.net

The Author, the publisher and all interested parties have used all possible care to assure that the information contained in this book is as complete and as accurate as possible. However, neither the publisher nor the author nor any interested party assumes any liability for omissions, errors, or defects in the materials, instructions or diagrams contained in this publication, and therefore are not liable for any damages including (but not limited to) personal injury, property damage or legal disputes that may result from the use of this book.

All major appliances are complex electro-mechanical devices. Personal injury or property damage may occur before, during, or after any attempt to repair an appliance. This publication is intended for individuals posessing an adequate background of technical experience. The above named parties are not responsible for an individual's judgement of his or her own technical abilities and experience, and therefore are not liable for any damages including (but not limited to) personal injury, property damage or legal disputes that may result from such judgements.

The advice and opinions offered by this publication are of a subjective nature ONLY and they are NOT to be construed as legal advice. The above named parties are not responsible for the interpretation or implementation of subjective advice and opinions, and therefore are not responsible for any damages including (but not limited to) personal injury, property damage, or legal disputes that may result from such interpretation or implementation.

The use of this publication acknowledges the understanding and acceptance of the above terms and conditions.

Special Thanks to technical consultant Dick Miller, whose forty years of experience as a field appliance service technician are reflected in the technical information and procedures described in this publication.

Published by EB Publishing, Inc., 251 Jeanell Drive Suite 3, Carson City, NV 89703

Printed in the United States of America

HOW TO USE THIS BOOK

STEP 1: READ THE DISCLAIMERS ON THE PREVIOUS PAGE. This book is intended for use by people who have a bit of mechanical experience or aptitude, and just need a little coaching when it comes to appliances. If you don't fit that category, don't use this book! We're all bloomin' lawyers these days, y'know? If you break something or hurt yourself, no one is responsible but **YOU**; not the author, the publisher, the guy or the store who sold you this book, or anyone else. Only **YOU** are responsible, and just by using this book, you're agreeing to that, and a lot more. If you don't understand the disclaimers, get a lawyer to translate them **before** you start working.

Read the safety and repair precautions in chapter 2. These should help you avoid making too many *really* bad mistakes.

STEP 2: READ CHAPTERS 1, 2 & 3: Know what kind of oven or cooktop, range or stove you have and basically how it works. When you go to the appliance parts dealer, have the nameplate information at hand. Have the proper tools at hand, and know how to use them.

STEP 3: READ CHAPTER 4, 5 OR 6 ABOUT YOUR COOKING EQUIPMENT.

STEP 4: FIX THE BLOOMIN' THING! If you can, of course. If you're just too confused, or if the book recommends calling a technician for a complex operation, call one.

WHAT THIS BOOK WILL DO FOR YOU
(and what it won't!)

This book **will** tell you how to fix the most common problems with the most common brands of domestic (household) cooking equipment. (This represents 95+ percent of all repairs that the average handyman or service tech will run into.)

This book **will not** tell you how to fix your industrial or commercial or any very large cooking equipment. The support and control systems for such units are usually very similar in function to those of smaller units, but vastly different in design, service and repair.

We **will** show you the easiest and/or fastest method of diagnosing and repairing your cooking equipment.

We **will not necessarily** show you the absolute cheapest way of doing something. Sometimes, when the cost of a part is just a few dollars, we advocate replacing the part rather than rebuilding it. We also sometimes advocate replacement of an inexpensive part, whether it's good or bad, as a simplified method of diagnosis or as a preventive measure.

We **will** use only the simplest of tools; tools that a well-equipped home mechanic is likely to have and to know how to use, including a VOM.

We **will not** advocate your buying several hundred dollars' worth of exotic equipment or special tools, or getting advanced technical training to make a one-time repair. It will usually cost you less to have a professional perform this type of repair. Such repairs represent only a very small percentage of all needed repairs.

We **do not** discuss electrical, mechanical or thermodynamic theories in detail. There are already many very well-written textbooks on these subjects and most of them are not likely to be pertinent to the job at hand; fixing your cooking equipment!

We **do** discuss rudimentary mechanical systems, heat flow, and simple electrical circuits.

We expect you to be able to look at a part and remove it if the mounting bolts and/or connections are obvious. If the mounting mechanism is complicated or hidden, or there are tricks to removing or installing something, we'll tell you about it.

You are expected to know what certain electrical and mechanical devices are, what they do in general, and how they work. For example, switches, relays, heater elements, motors, solenoids, cams, gas valves, air seals, and centrifugal blowers. If you do not know what these things do, learn them BEFORE you start working on your cooking equipment.

You should know how to cut, strip, and splice wire with crimp-on connectors, wire nuts and electrical tape. You should know how to measure voltage and how to test for continuity with a VOM (Volt-Ohm Meter). If you have an ammeter, you should know how and where to measure the current in amps. If you don't know how to use these meters, there's a brief course on how to use them (for *our* purposes *only*) in Chapter 2. Read chapter 2 before you buy either or both of these meters.

A given procedure was only included in this book if it passed the following criteria:

1) The job is something that the average couch potato can complete in one afternoon, with no prior knowledge of the appliance, with tools a normal home handyman is likely to have.
2) The problem is a common one; occuring more frequently than just one out of a hundred appliances.
3) The parts and/or special tools required to complete the job are easily found and not too expensive; and the cost of the repair is far less than replacing the appliance or calling a professional service technician.
4) The repair is likely to yield an appliance that will operate satisfactorily for several more years, or at least long enough to justify the cost.

I'm sure that a physicist reading this book could have a lot of fun tearing it apart because of my deliberate avoidance and misuse of technical terms. However, this manual is written to simplify the material and inform the novice, not to appease the scientist.

__NOTE:__ The diagnosis and repair procedures in this manual do not necessarily apply to brand-new units, newly-installed units or recently relocated units. Although they __may__ posess the problems described in this manual, cooking equipment that has recently been installed or moved are subject to special considerations not taken into account in this manual for the sake of simplicity. Such special considerations include installation parameters, installation location, the possibility of manufacturing or construction defects, damage in transit, and others.

This manual was designed to assist the novice technician in the repair of household (domestic) cooking equipment that has been operating successfully for an extended period of months or years and has only recently stopped operating properly, with no major change in installation parameters or location.

Table Of Contents

Chapter 1

SYSTEM BASICS

1-1 BASIC FUNCTIONS

In their most basic forms, ovens and cooktops are pretty simple devices. Technically, all they do is develop high, controlled temperatures in specific places, in order to transfer heat to food and cook it.

The important words here are "high" and "controlled." These two requirements present specific challenges to the folks who design and service ovens. And consumers demand gadgets that might make life easier for them, but they sure make the life of a serviceman tougher.

For example, when you talk about controlled temperatures, you are not only talking about keeping the temperature within a certain range with a thermostat, but also controlling the *times* when the heat starts and stops being applied to the food. So there might be a timer wired into the oven circuit to start and stop it. In some models there is a self-cleaning function too, which means that the oven temperature will rise to an extraordinarily high level (around 1000 degrees), stay there for a set amount of time, and then shut off. You need special controls and safety mechanisms to deal with those temperatures.

The other important word here is "high." You need to use powerful energy sources to develop high temperatures. If these energy sources are not tightly controlled, the result can literally be a disaster. For example, before you open a gas valve to a burner, you need to make sure the ignition source is working. If it isn't, you certainly don't want that gas valve to open and dump a bunch of unburned gas into your oven or kitchen. Talk about an explosion hazard! So you design a safety mechanism to prevent it.

One of the most amazing collections I've ever seen resides at Appliance Parts Equipment Company in Santa Rosa, California. It is a bunch of electric cooktop surface units and gas burner grates that have been brought in over the years, with various things *melted* onto them, from kids' toys to glass plates to aluminum tea kettles, and even an aluminum pressure cooker. If seeing this collection doesn't give you some respect for the heat and power you're dealing with in cooking equipment, nothing will. I suppose there's some profound philosophical lesson in this collection about man harnessing the forces of nature or something, but metaphysics are a little beyond the scope of this manual, so let's move on....

1-2 ELECTRIC COOKTOPS AND OVENS

Most electric cooking equipment uses two different electrical circuits. The heating elements usually run on 220 volts, and accessories such as lights, timers and rotisserie motors run on 110 volts. There are a few notable exceptions. Some smaller "apartment" cooktops run on 110 volts. Also, in some fixed-temperature switch applications, 110 volts is applied to a 220 volt surface unit (burner) to achieve a "low" heat setting.

In most cooktops, the heating element is simply a big resistor wire, with enough resistance to generate a high heat. Usually these are nichrome wire, surrounded in ceramic insulation, with a steel sheath around the ceramic. On higher settings, the element glows red when operating. Heating occurs mainly by conduction; that is, the direct contact of the heating element to the cookware. Since the surface unit coil is flat, flat-bottomed cookware provides the best contact with these units and thus the most efficient operation.

A fairly recent development is the radiant heat cooktop. These have a radiant element (something like a very intense sunlamp) underneath a glass surface. These units do not heat the pot or pan by direct contact (conduction) like coil surface units. They heat by radiation, much like a sunlamp heats your skin.

To maintain a set temperature, the element is cycled on and off, usually by a switch called an *infinite switch*, so named because it theoretically provides an infinite number of heat settings. There are also *fixed-temperature swit-ches* that vary the voltage going to the heating elements to maintain fixed, pre-set temperatures. These are push-button or rotary switches with fixed settings such as warm, low, medium and high. These switches and systems are discussed in detail in chapter 4.

1-3 GAS COOKTOPS AND OVENS

Gas ovens use the burning of a fuel such as natural gas, LPG or propane to generate heat. These gases are obviously highly flammable and are heavier than air; they must be closely controlled to prevent explosion hazards. The different fuels require valves and burners with different orifice sizes, so when buying parts, make sure you get the right ones for the fuel you are using.

A pressure regulator keeps the gas entering the stove at a constant pressure of about 1/6 PSI, regardless of fluctuations in the supply pressure. In a cooktop or stove, this pressure regulator feeds a main gas header, or *manifold*, located under the cooktop. The surface burner gas valves are mounted directly to the gas header. Gas is piped from the header to the various burners, pilots and safety valves, and in some systems, the oven thermostat.

Temperature control in cooktops is very different from that in ovens. In cooktops, a gas valve varies the flow of gas to the burner. In ovens, the gas is either on or off; the burner cycles on and off to maintain temperature.

Another major difference is that when you turn on a gas cooktop, you can immediately see if it ignites. If it doesn't, you turn off the burner and figure out why. In ovens, since the burner is inside the oven, you cannot immediately see whether the burner has ignited. If it does not ignite, you certainly do not want the gas valve to stay open. This would dump raw unburned gas into the oven and create an explosion hazard.

This creates different ignition and safety needs for cooktops versus ovens. Cooktops use a standing pilot or spark ignition system. Ovens use a standing pilot, spark or glow bar ignition system, and gas safety valves that will not open unless ignition is assured. These systems are discussed in detail in chapters 5 & 6.

1-4 SELF CLEANING OVENS

In a self-cleaning oven, the oven temperature will rise to an extraordinarily high level (around 1000 degrees), stay there for a set amount of time (usually 2-3 hours) while the heat incinerates everything, and then shut off. These ovens require a timer to control the length of the cleaning cycle. Of course, you do not want the temperature to go much above 1000 degrees, so you need extra thermostatic controls and fail-safe mechanisms to prevent that.

Neither do you want the user to open the oven door in the middle of the cleaning cycle and get a face full of 1000 degree air. At that temperature there is also a risk of flashback, where the oven temperature is so high that it flashes (burns) the oxygen right out of the air in your kitchen. An oven door locking mechanism prevents the door from being opened at high temperatures. The self-cleaning function is electrically interlocked with the door locking mechanism, so the self-clean-

ing function cannot be used unless the door locking system is engaged.

A door locking system has two steps. First there is a manual latch that must be engaged to signal your "intent to clean." Then when the oven temperature climbs above about 550 degrees, an automatic doorlock system engages and prevents manual opening until the oven temperature drops back down below 550. The automatic doorlock might be a bimetal or a solenoid, or in some machines there is a very low RPM electric motor (like 1 RPM) that rotates. As a solenoid or motor locking system engages, it throws switches that prevent it from being energized again until a cool-down thermostat tells it that oven conditions are safe to do so.

The temperature of the self-cleaning cycle is controlled by one of two different means. Either the main oven thermostat has a second sensor built into it, or there is a separate (cleaning) thermostat altogether, with its own temperature probe. The probe usually sticks into the oven through the back or side wall, near the top.

A cleaning cycle will always be wired through a timer. In some machines there is a fixed-time cleaning cycle, but usually you can set the cleaning cycle for however long you want it to last. Usually 2-3 hours is what's recommended.

Self-cleaners tend to be very complex machines. They have many extra safety mechanisms to prevent overheating and burns to the user. When diagnosing these things, you *must* use a wiring diagram. That's the only way to know for sure what switches and interlocks are used by a particular system. See chapter 2 for details.

1-5 CONVECTION OVENS

In order to understand convection ovens, there are a few principles you need to understand first:

1) Ovens *don't* make things hot. They *add heat* to whatever you put into them.

2) Heat will always flow *from* something of a *higher* temperature *to* something of a *lower* temperature. The farther apart the temperatures are, the faster the heat flow.

3) Heat will continue to flow from one object to another until the temperatures of the two objects are equal.

4) Air is really a poor conductor of heat. It is actually a pretty good insulator.

Lets talk about chill factor for a minute. Chill factor? Isn't that *weather* stuff? In an oven manual? Yeah, because the concepts are the same. Stick with me here.

If the weather outside is freezing, this may sound funny, but it does not make you cold. What it does is to *remove heat* from your body, and heat flowing from your body into the air around you is what makes you feel cold.

If the wind is not blowing, your body transfers heat to the cold air around it. The temperature of the air closest to your skin starts to rise. Soon you have a little "blanket" of relatively warm air around you. As the air temperature of this "blanket" rises, the heat flow from your body slows down. When the heat flow slows down, you don't feel as cold.

If the wind is blowing, the air touching your skin does not have a chance to warm. The wind is constantly blowing away the warm layer of air and replacing it with cold air. The heat flow does not slow down, and you feel colder, *even though the outside air temperature is the same as before*. We humans refer to this as the *chill factor*.

The same thing happens inside an oven. If the air is still, the heat does not flow from the air to the food as fast, and cooking is slower. If the air is moving, heat gets transferred faster and cooking occurs faster.

All ovens have *some* air moving around inside, due to natural convection (warm air rises, cooler air falls.) In a *convection* oven, a fan is used to force the air to move around inside the oven, speeding up the cooking process.

The fan also has two other functions in the oven. Oven temperatures are pretty extreme conditions in which to operate an electric motor. If the oven is also self-cleaning the temperatures are even higher. So the fan motor actually draws air at room-temperature over itself to keep itself cool.

In a gas oven, air is also needed for combustion. The fan pushes air through the burner. In this system there will also be something called a "sail" switch. This is a switch with a little metal "sail" that activates it. Air from the fan closes the switch contacts when the fan is operating. The sail switch is wired in series with the heating system of a convection oven. If the blower fan is not operating, it is not cooling itself, so you do not want the heating system on. In addition, in a *gas* convection oven, proper combustion will not occur without proper airflow.

1-6 MEAT PROBES

Some oven models were built with meat probes. A meat probe is simply a combination thermometer and variable resistor, called a thermistor, that sticks into meat that you're cooking. The resistance of the thermistor varies with the temperature inside the meat. Then when the meat reaches a certain internal temperature, the thermistor reaches a certain resistance, and a buzzer sounds, or the oven cycles on and off to maintain temperature.

To lessen the hazard of shock, usually meat probes are on a separate low voltage (12 volt) control circuit. To get 12 volts from the 110 volt power supply, a transformer is needed, as well as additional circuitry for the shutdown function, etc.

1-7 ROTISSERIES

A rotisserie is just a slow-turning 110 volt motor that turns meat over as you cook it. Getting a small, low-horse-power motor to turn over ten unbalanced pounds of roast beast requires some mechanical help, so the motor drives the spit through a gear train.

Chapter 2

TROUBLESHOOTING TOOLS AND SAFETY TIPS AND TRICKS

Lets talk aluminum foil for a minute. If you line the bottom of your oven with it, it can block the airflow within your oven. If you have a convection oven, it *really* defeats the purpose. But even in an oven with natural convection, it can mess up airflow and cooking and even cause burners to malfunction. If you simply *must* line the bottom of your oven with foil, at least poke holes in it where there are holes in the oven floor. They're there for a reason.

2-1 BASIC REPAIR AND SAFETY PRECAUTIONS

1) When working on gas cooking equipment, if you've disconnected a gas pipe to replace a valve or other component, always test the pipe joint for leaks when you reassemble it. You can do this by coating the joint with a solution of liquid soap and water and looking for bubbles. Apply it with a brush to make sure you coat the joint thoroughly, and use a mirror to look at the back side of the joint if necessary. Your appliance parts dealer has gas leak testing solution, with a brush built into the cap, made specifically for this purpose.

2) Always de-energize (pull the plug or trip the breaker on) any oven that you're disas-

sembling. If you need to re-energize the oven to perform a test, make sure any bare wires or terminals are taped or insulated. Energize the unit only long enough to perform whatever test you're performing, then disconnect the power again.

3) If this manual advocates replacing the part, REPLACE IT!! You might find, say, a solenoid that has jammed for no apparent reason. Sometimes you can clean it out and lubricate it, and get it going again. The key words here are *apparent reason*. There is a reason that it stopped—you can bet on it— and if you get it going and re-install it, you are running a very high risk that it will fail again. If that happens, you will have to start repairing your oven all over again. It may only act up when it is hot, or it may be bent slightly...there are a hundred different "what if's." Very few of the parts mentioned in this book will cost you over ten or twenty dollars. Replace the part.

4) Always replace the green (ground) leads when you remove an electrical component. They're there for a reason. And NEVER EVER remove the third (ground) prong in the main power plug! They are *especially* important in spark ignition systems; the spark electrode will not spark without grounding.

5) When opening the oven cabinet or console, remember that the sheet metal parts have very sharp edges. Wear gloves, and be careful not to cut your hands!

6) If you have diagnosed a certain part to be bad, but you cannot figure out how to remove it, sometimes it helps to get the new part and examine it for mounting holes or other clues as to how it may be mounted.

7) When testing for a 110 volt power supply from a wall outlet, you can plug in a small appliance such as a shaver or blow dryer. If you're not getting full power out of the outlet, you'll know it right away. If you're testing for 220 volt power you need to use the VOM.

8) When splicing wires in an oven, remember that you're dealing with high temperatures. Normal connectors and wire insulation will melt under these conditions. Your parts dealer has high-temp connections, porcelain wire nuts and fiberglass-insulated wire for this purpose.

I want to impress upon you something really important. In electric cooking equipment, you're usually dealing with 220 volt circuits. DO NOT TAKE THIS LIGHTLY. I've been hit with 110 volts now and then. Anyone who works with electrical equipment has at one time or another. It's unpleasant, but unless exposure is more than a second or so, the only harm it usually does is to tick you off pretty good. However, *220 VOLTS CAN KNOCK YOU OFF YOUR FEET. IT CAN DO YOUR BODY SOME SERIOUS DAMAGE, VERY QUICKLY. DO NOT TEST LIVE 220 VOLT CIRCUITS.* If you have a heart condition, epilepsy, or other potentially serious health conditions, well...hey, it's just my opinion, but you shouldn't be testing 220 volt circuits *at all*. It's not worth dying for.

2-2 BEFORE YOU START

Find yourself a good appliance parts dealer. You can find them in the yellow pages under the following headings:

● Appliances, Household, Major
● Appliances, Parts and Supplies
● Refrigerators, Domestic
● Appliances, Household, Repair and Service

Call a few of them and ask if they are a repair service, or if they sell parts, or both. Ask them if they offer free advice with the parts they sell. (Occasionally, stores that offer both parts and service will not want to give you advice.) Often the parts counter men are ex-technicians who got tired of the pressures of in-home service. They can be your best friends. However, you don't want to badger them with too many questions, so know your basics before you start asking questions.

Some parts houses may offer service, too. Be careful! There may be a conflict of interest. They may try to talk you out of even trying to fix your own oven. They'll tell you it's too complicated, then in the same breath "guide" you to their service department. Who are you gonna believe, me or them? Not all service and parts places are this way, however. If they genuinely try to help you fix it yourself, and you find that you're unable to, they may be the best place to look for service. Here's a hot tip: after what I just said, if they sold you this book, then I'll just about guarantee they're genuinely interested in helping do-it-yourselfers.

When you go into the store, have ready the make, model and serial number from the nameplate of the oven. Chapter 3 has information on how to find the nameplate. On some models, you will also need the *lot number* to get the right part, so if there is one on the nameplate, write that down, too.

2-3 TOOLS (Figure 2-A)

Most of the tools that you might need are shown below. Some are optional. The reason for the option is explained.

ELECTRICAL PLIERS or STRIPPERS and DIAGONAL CUTTING PLIERS: For cutting and stripping small electrical wire

VOM (VOLT-OHM METER): For testing electrical circuits. If you do not have one, get one. An inexpensive one will suffice, as long as it has both "AC Voltage" and "Resistance" (i.e. Rx1, Rx10) settings on the dial. It will do for our purposes.

BUTT CONNECTORS, CRIMPERS, WIRE NUTS and ELECTRICAL TAPE: For splicing small wire. When splicing wire on ovens, you must make sure you use high-temp connections, such as porcelain wire nuts, high-temp terminals and fiberglass insulated wire.

ALLIGATOR JUMPERS (sometimes called a "CHEATER" or "CHEATER WIRE":) Small gauge (14-16 gauge or so) and about 12-18 inches long, for testing electrical circuits. Available at your local electronics store. Cost: a few bucks for 4 or 5 of them.

SCREWDRIVERS: Both flat and Phillips head; two or three sizes of each. It's best to have at least a stubby, 4- and 6-inch sizes.

NUTDRIVERS: You will need at least 1/4" and 5/16" sizes. 4- and 6-inch ones should suffice, but it's better to have a stubby, too.

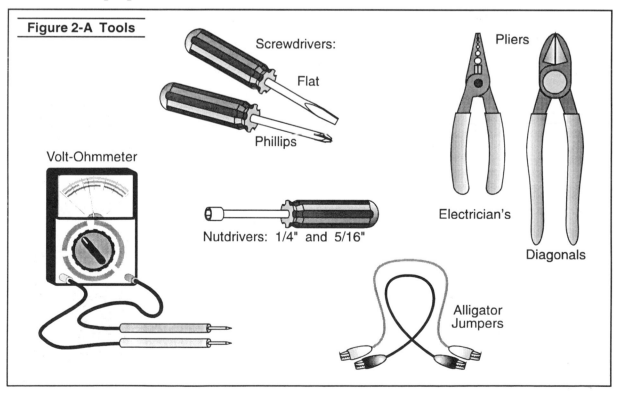

Figure 2-A Tools
Screwdrivers:
Flat
Phillips
Pliers
Volt-Ohmmeter
Nutdrivers: 1/4" and 5/16"
Electrician's
Diagonals
Alligator Jumpers

OPTIONAL TOOLS *(Figure 2-B)*

SNAP-AROUND AMMETER: For determining if electrical components are energized, without cutting into the system. Quite useful; but a bit expensive, and there are alternate methods. If you have one, use it; otherwise, don't bother getting one.

EXTENDIBLE INSPECTION MIRROR: For seeing difficult places beneath the oven and behind panels.

CORDLESS POWER SCREWDRIVER OR DRILL/DRIVER WITH MAGNETIC SCREWDRIVER AND NUTDRIVER TIPS: For pulling off panels held in place by many screws. It can save you lots of time and hassle.

2-4 HOW TO USE A VOM AND AMMETER

Many home handymen are very intimidated by electricity. It's true that diagnosing and repairing electrical circuits requires a bit more care than most operations, due to the danger of getting shocked. But there is no mystery or voodoo about the things we'll be doing. Remember the rule in section 2-1; while you are working on a circuit, energize the circuit only long enough to perform whatever test you're performing, then take the power back off it to perform the repair. You need not be concerned with any theory, like what an ohm is, or what a volt is. You will only need to be able to set the VOM onto the right scale, touch the test leads to the right place and read the meter.

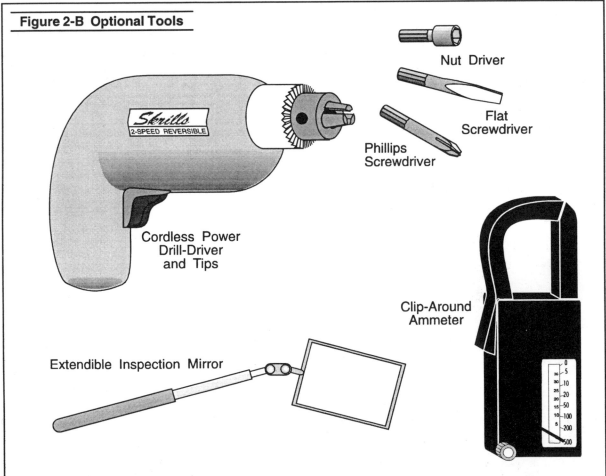

Figure 2-B Optional Tools

Nut Driver
Flat Screwdriver
Phillips Screwdriver
Cordless Power Drill-Driver and Tips
Clip-Around Ammeter
Extendible Inspection Mirror

In using the VOM (Volt-Ohm Meter) for our purposes, the two test leads are always plugged into the "+" and "-" holes on the VOM. (Some VOMs have more than two holes.)

2-4(a) TESTING VOLTAGE (Figure 2-C)

Set the dial of the VOM on the lowest VAC scale (A.C. Voltage) over 120 volts. For example, if there's a 50 setting and a 250 setting on the VAC dial, use the 250 scale, because 250 is the lowest setting over 120 volts.

If you're testing 220 volt circuits, use the lowest scale over 220 volts.

Touch the two test leads to the two metal contacts of a live power source, like a wall outlet or the terminals of the motor that you're testing for voltage. (Do not jam the test leads into a wall outlet!) If you are getting power through the VOM, the meter will jump up and steady on a reading. You may have to convert the scale in your head. For example, if you're using the 250 volt dial setting and the meter has a "25" scale, simply divide by 10; 120 volts would be "12" on the meter.

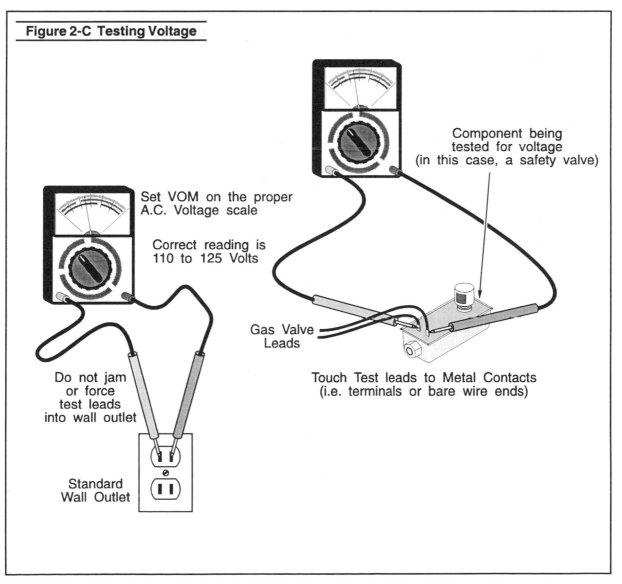

Figure 2-C Testing Voltage

Set VOM on the proper A.C. Voltage scale

Correct reading is 110 to 125 Volts

Component being tested for voltage (in this case, a safety valve)

Gas Valve Leads

Do not jam or force test leads into wall outlet

Touch Test leads to Metal Contacts (i.e. terminals or bare wire ends)

Standard Wall Outlet

2-4(b) TESTING FOR CONTINUITY (Figure 2-D)

Don't let the word "continuity" scare you. It's derived from the word "continuous." In an electrical circuit, electricity has to flow from a power source back to that power source. If there is any break in the circuit, it is not continuous, and it has no continuity. "Good" continuity means that there is no break in the circuit.

For example, if you were testing an ignitor to see if it was burned out, you would try putting a small amount of power through the ignitor. If it was burned out, there would be a break in the circuit, the electricity wouldn't flow, and your meter would show no continuity.

That is what the resistance part of your VOM does; it provides a small electrical current (using batteries within the VOM) and measures how fast the current is flowing. For our purposes, it doesn't matter how *fast* the current is flowing; only that there *is* current flow.

To use your VOM to test continuity, set the dial on (resistance) R x 1, or whatever the lowest setting is. Touch the metal parts of the test leads together and read the meter. It should peg the meter all the way on the right side of the scale, towards "0" on the meter's "resistance" or "ohms" scale. If the meter does not read zero ohms, adjust the thumbwheel on the front of the VOM until it does read zero. If you cannot get the meter to read zero, the battery in the VOM is low; replace it.

If you are testing, say, an ignitor, first make sure that the burner leads are not connected to anything, especially a power source. If the ignitor's leads are still connected to something, you may get a reading through that something. If there is still live power on the item you're testing for continuity, you will burn out your VOM instantly and possibly shock yourself.

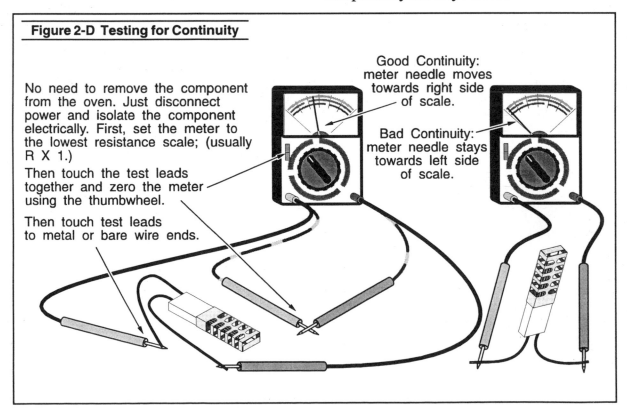

Figure 2-D Testing for Continuity

No need to remove the component from the oven. Just disconnect power and isolate the component electrically. First, set the meter to the lowest resistance scale; (usually R X 1.)

Then touch the test leads together and zero the meter using the thumbwheel.

Then touch test leads to metal or bare wire ends.

Good Continuity: meter needle moves towards right side of scale.

Bad Continuity: meter needle stays towards left side of scale.

Touch the two test leads to the two bare wire ends or terminals of the ignitor. You can touch the ends of the wires and test leads with your hands if necessary to get better contact. The voltage that the VOM batteries put out is very low, and you will not be shocked. If there is NO continuity, the meter won't move. If there is GOOD continuity, the meter will move toward the right side of the scale and steady on a reading. This is the resistance reading and it doesn't concern us; we only care that we show good continuity. If the meter moves only very little and stays towards the left side of the scale, that's BAD continuity; the ignitor is no good.

If you are testing a switch, you will show little or no resistance (good continuity) when the switch is closed, and NO continuity when the switch is open. If you do not, the switch is bad.

2-4(c) AMMETERS

Ammeters are a little bit more complex to explain without going into a lot of electrical theory. If you own an ammeter, you probably already know how to use it.

If you don't, don't get one. Ammeters are expensive. And for our purposes, there are other ways to determine what an ammeter tests for. If you don't own one, skip this section.

For our purposes, ammeters are simply a way of testing for continuity without having to cut into the system or to disconnect power from whatever it is we're testing.

Ammeters measure the current in amps flowing through a wire. The greater the current that's flowing through a wire, the greater the density of the magnetic field, or flux, it produces around the wire. The ammeter simply measures the density of this flux, and thus

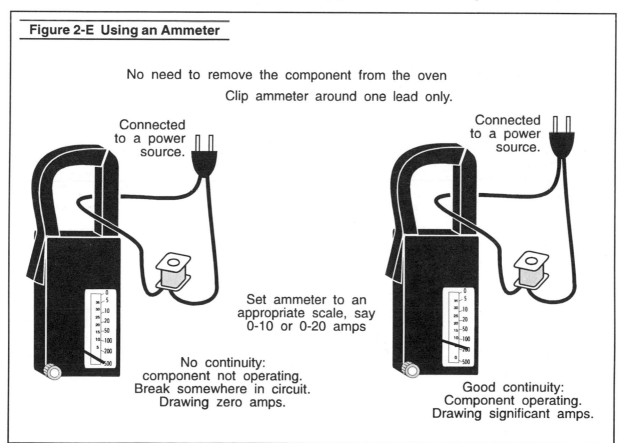

Figure 2-E Using an Ammeter

No need to remove the component from the oven
Clip ammeter around one lead only.

Connected to a power source.

Connected to a power source.

Set ammeter to an appropriate scale, say 0-10 or 0-20 amps

No continuity: component not operating. Break somewhere in circuit. Drawing zero amps.

Good continuity: Component operating. Drawing significant amps.

the amount of current, flowing through the wire. To determine continuity, for our purposes, we can simply isolate the component that we're testing (so we do not accidentally measure the current going through any other components) and see if there's *any* current flow.

To use your ammeter, first make sure that it's on an appropriate scale (0 to 10 or 20 amps will do). Isolate a wire leading directly to the component you're testing. Put the ammeter loop around that wire and read the meter. (Figure 2-E)

For example, let's say you're trying to tell if the oven is using any current anywhere. Clamp the ammeter around just one of your main wall power leads. If the meter shows any reading at all, *something* in the oven is using power.

2-5 WIRING DIAGRAM

Sometimes you need to read a wiring diagram, to make sure you are not forgetting to check something. Sometimes you just need to find out what color wire to look for to test a component. It is ESPECIALLY important in diagnosing self-cleaning ovens.

Usually your wiring diagram is either pasted to the back of the oven, or else contained in a plastic pouch inside the backguard or inside the leg of the oven, near the broiler door (see section 3-6)

If you already know how to read a wiring diagram, you can skip this section.

Each component should be labelled clearly on your diagram. Look at figure 2-F. The symbols used to represent each component are pretty universal.

Wire colors are abbreviated and shown next to each wire. For example, Y means a yellow wire, PK means pink, R means red. Black is usually abbreviated BK, blue is usually BU. GR or GN are green, GY is gray. A wire color with a dash or a slash means --- with a --- stripe. For example BU-W means blue with a white stripe, T/R means tan with a red stripe.

A few notes about reading a wiring diagram:

Notice that in some parts of the diagram, the lines are inside a dashed box. These switches and wiring are inside of the timer or other block of switches. In some wiring diagrams, wiring and switches inside a timer or other switchblocks are drawn with lines that are thicker than the rest of the wiring.

The small white circles all over the diagram are terminals. These are places where you can disconnect the wire from the component for testing purposes. The small black circles indicate places where one wire is connected to the other. If two wires cross on the diagram without a black dot, they are not connected.

Switches may be numbered or lettered. Usually the terminals on the outside of the timer or switch are stamped or printed with markings that you will see on the wiring diagram.

To test a switch, mark and disconnect all the wires. Connect your ohmmeter to the two terminal leads of the switch you want to test. For example, in figure 2-F, if you want to test the door switch, take power off the machine, disconnect the wires from it and connect one test lead to COM and one to NC. Then flick the switch back and forth. It should close and open. If it does, you know that contact inside the switch is good.

Remember that for something to be energized, it must make a complete electrical circuit. You must be able to trace the path that the electricity will take, FROM the wall outlet back TO the wall outlet. This includes not only the component that you suspect, but all switches leading to it, and sometimes other components, too.

Figure 2-F Wiring Diagram

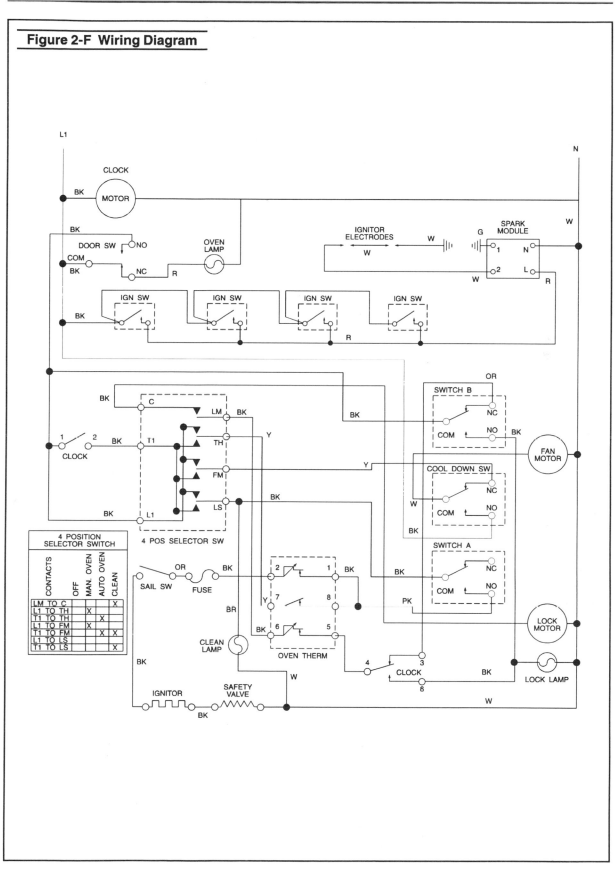

110 VOLTS VS 220 VOLTS: ELECTRIC OVENS VS GAS OVENS

Electric circuits in *gas* ovens are 110 volt circuits. Most *electric* cooking equipment has two circuits; 110 volt and 220 volt. (see figure 2-G) In a 220 volt circuit, L1 to L2 will measure 220 volts, but L1 to N or L2 to N will measure 110 volts. In this simplified circuit diagram, notice that only the heating elements operate on 220 volts. The timer, lights and other accessories operate on 110.

Let's trace an actual cleaning circuit. Figure 2-H shows a pretty complex wiring diagram. It is for a self-cleaning gas convection oven, with spark surface ignitors. When tracing an actual circuit, especially in more complex diagrams, it may be worthwhile to make a couple of photocopies of the wiring diagram, and trace circuits with a colored highlighter pen, to keep yourself straight.

Following the gray-shaded circuit, note that the electricity "flows" from L1 to N. This tells you that it is a 110 volt circuit.

Let's say the oven door is not locking during the cleaning cycle. Since the locking mechanism is interlocked with the heating circuit, the oven will not reach cleaning temperature either.

Let's start at the lock motor and find out which switches feed electricity to it. One of the leads goes directly to N. Tracing the other lock motor lead, we first end up at the "C" terminal of the 4-position selector switch. Looking at the switch for the chart, the "C" to "LM" contacts are closed when the "clean" button is depressed on the switch. So we need to test that switch for proper operation. (see section 2-6(a))

Continuing with the circuit, we leave the selector switch through the "LM" terminal, and enter the oven thermostat through the "6" terminal. Power then leaves the thermostat through the "5" terminal, so we need to check for continuity between terminals "5" and "6" of the thermostat. Since it is a thermostatic switch, only heat will open the switch, so we only need to test it for continuity. If it is not opening when it gets to temperature, that will cause different symptoms from what we're trying to diagnose.

Likewise terminals "4" and "3" of the clock (timer.) Make sure the switch is closed and test for continuity between those terminals.

Switch B, switch A, and the cool-down switch are all activated when the locking motor turns. They are shown in the diagram in their "normal state." So continuing with our circuit, if the locking motor is not turning, you need to check switch "B" for continuity between the "COM" and "NC" terminals. If there is no continuity, it might mean the switch is bad. *It might also mean that the switch was not returned to its "normal" state the last time it*

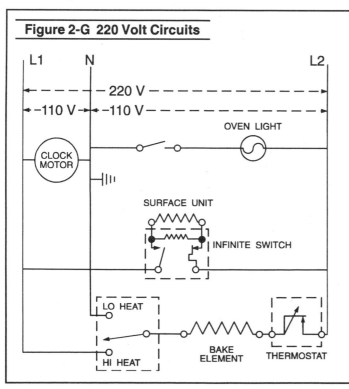

Figure 2-G 220 Volt Circuits

was activated! You need to examine the switch carefully to determine what the problem is. If the locking motor stopped turning before the switch unlocked, you've got other problems. You will need to trace other circuits in the diagram to figure out what.

From the "COM" terminal of switch "B," the circuit goes back through the door switch to L1. The door switch feeds several other oven circuits, too, so unless there's something else not working, we can eliminate that as the problem. Note that in its "normal" state, the oven light is on and the door is open. The door must be closed to close the switch that feeds electricity to the thermostats and heating circuit.

To check for a wire break, you would pull each end of a wire off the component and test for continuity through the wire. You may need to use jumpers to extend or even bypass the wire; for example, if one end of the wire is in the control console and the other end in underneath the machine. If there is no continuity, there is a break in the wire! It will then be up to you to figure out exactly where that break is—there is no magic way. If you have a broken wire, look along the length of the wire for pinching or chafing. The most likely place you will find burnt wires is inside the back panel. If there is a place where the wires move, check there too. Even if the insulation is O.K., the wire may be broken inside.

Figure 2-H Tracing a Wiring Diagram

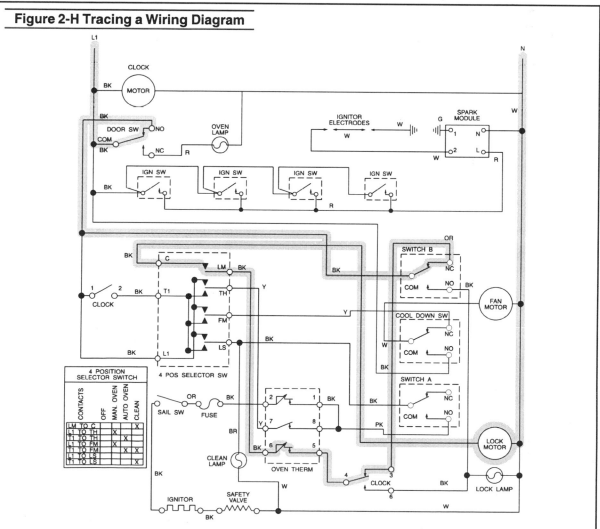

2-6 TESTING COMPONENTS

Most components are tested simply by removing power and placing a resistance meter across them. However some need to be tested with a VOM while energized. An ammeter is a safer way to test energized circuits if you have one, especially testing 220 volt circuits. Occasionally, if the component is inexpensive enough, it's easier to just replace it and see if that solves the problem.

Following is a primer on how to test each individual component you might find in any given oven. Diagnostic information is covered in chapters 4, 5 & 6.

2-6(a) SWITCHES

Testing switches and solenoids is pretty straightforward. Take all wires off the component and test continuity across it as described in section 2-4(b).

Switches should show good continuity when closed and no continuity when open. Flick the switch back and forth with the resistance meter attached and see if the contacts are opening and closing.

NC means "normally closed;" in other words, if nothing is touching the switch, you should see continuity between the terminals. NO means "normally open" and with the switch at rest, you will see no continuity through it.

SELECTOR SWITCHBLOCKS

A selector switchblock, located in the control panel, is a group of switches all molded into one housing. In your oven, a switchblock might be used to allow you to choose a cooktop heat setting, for example, or a cleaning cycle instead of a baking cycle. When you *select* an option, "high heat" for example, you are *de-selecting* the other options, for example "medium heat." This is why you would use a switchblock instead of individual switches.

Testing switch blocks is much like testing timers. You must look at the wiring diagram to see which of the terminals will be connected when the internal switches are closed.

Keep in mind, however, that you must also know which of the internal switches close when an external button is pressed. When you press one button on the switchblock, *several* of the switches inside may close at once. To test a switchblock, in addition to the

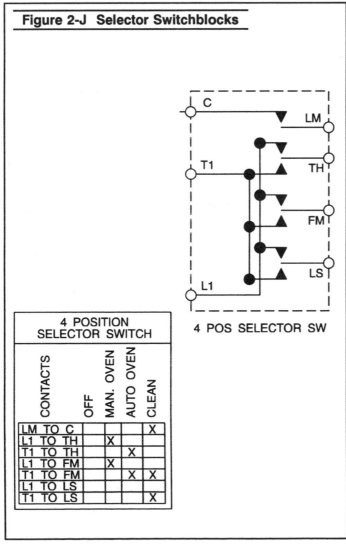

Figure 2-J   Selector Switchblocks

4 POS SELECTOR SW

4 POSITION SELECTOR SWITCH				
CONTACTS	OFF	MAN. OVEN	AUTO OVEN	CLEAN
LM TO C				X
L1 TO TH	X			
T1 TO TH		X		
L1 TO FM	X			
T1 TO FM		X	X	
L1 TO LS				
T1 TO LS				X

wiring diagram, you must have a chart that gives you this info. Usually this is a part of the wiring diagram.

Using the diagram and chart in figure 2-J, lets say we want to test contacts "T1" to "FM" for proper operation. We see that with the "clean" button pressed, these contacts inside the switchblock should be closed. With the "off" button pressed, these contacts should be open. Remove the wires and put the resistance meter probes on those two terminals. Then push the "off" and "clean" buttons alternately, to see if the switch opens and closes.

Test other switches similarly. Figure out when they're closed, when they're open, and test resistance while operating the switch.

If the switchblock is bad, replace it.

2-6 (b) OVEN CONTROLS - GENERAL

Analog, or electro-mechanical controls are covered thoroughly in this service manual. Solid-state digital controls, clocks and thermostats are not. Although error (fault) codes for most major brands may be found in Chapter 7, often diagnosis consists of simply replacing circuit components until you find the bad one.

Usually the problem turns out to be a bad oven sensor, stuck or defective keypad, ERC (clock) unit or a circuit board. Sensors are not bad, but the other parts are expensive and electrical. Electrical parts are usually non-returnable, so you could go through a lot of money trying to solve the problem. If you have a problem that you think you've traced to a solid-state circuit, try Chapter 7, but if you are in any way unsure

of your diagnosis, call a factory-authorized technician.

Do not confuse this with electro-mechanical "digital" clocks, in which the "digits" are printed on a wheel that turns. These can be tested and rebuilt or replaced as described in this chapter.

2-6 (c) THERMOSTATIC CONTROLS

A thermostat is simply a switch or a gas valve that opens and closes according to the temperature it senses. There might be several different kinds of thermostatic switches in any given oven; high-limit stats to prevent overheating, cleaning thermostats and cool-down stats for the oven cleaning cycle, etc.

In an oven, the main control thermostat body is located in the main control panel, (see chapter 3) but there is a liquid-filled temperature sensing bulb that extends down into the oven. (see figure 2-K) Heat in the oven increases the pressure in the bulb, and a capillary carries the pressure back to the body of the thermostat, where it opens and closes switch contacts or a gas valve.

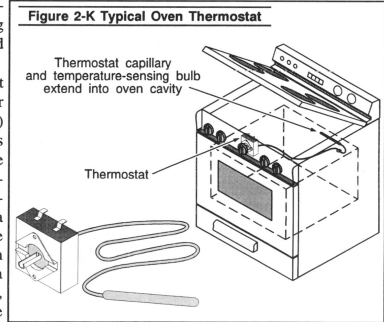

Figure 2-K Typical Oven Thermostat

Thermostat capillary and temperature-sensing bulb extend into oven cavity

Thermostat

Thermostats are either gas or electric; pressure from the capillary either closes an electrical switch or opens a gas valve.

That is the simplest form of a main control thermostat. If only all of them were that easy. If there is an automatic oven cycle, main control thermostats must also be wired through the timer. And if there is a cleaning cycle, they must be either bypassed or adjusted for the higher temperatures of that cycle. That's where you start to get dual-control thermostats, thermostats with twelve leads, and other complexities. Some control a pilot in addition to the main burner. Some even control two levels of the same pilot.

Main control thermostats are about the most expensive commonly-replaced parts in an oven. Usually the first thing a novice thinks is that the thermostat has quit working. It *should* be the *last* thing you conclude, after you have checked out everything else in the system.

The liquid inside the bulb and capillary of an oven thermostat is usually a mercury or sodium compound or some other such nasty and dangerous stuff. It can literally explode on contact with air. So when you replace an oven thermostat, do not cut open the capillary or bulb, and dispose of the old thermostat properly. The definition of "properly" varies between jurisdictions, but check with your appliance parts dealer or local fire department hazardous materials professionals.

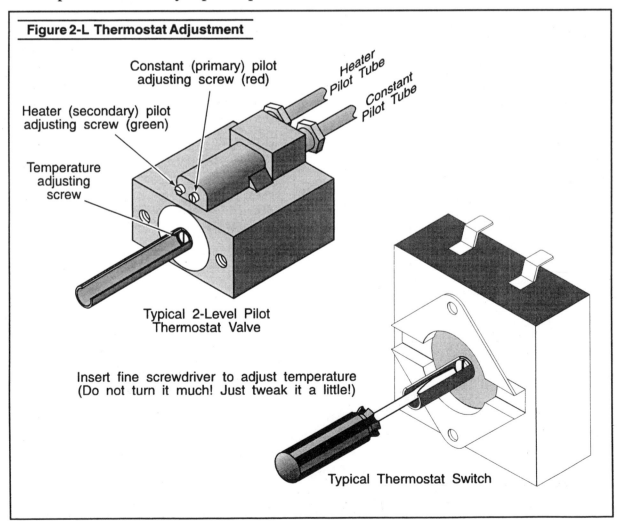

Figure 2-L Thermostat Adjustment

Constant (primary) pilot adjusting screw (red)

Heater Pilot Tube

Constant Pilot Tube

Heater (secondary) pilot adjusting screw (green)

Temperature adjusting screw

Typical 2-Level Pilot Thermostat Valve

Insert fine screwdriver to adjust temperature (Do not turn it much! Just tweak it a little!)

Typical Thermostat Switch

ADJUSTING THERMOSTATS

The oven is adjusted by a small adjusting screw in the center of the oven thermostat valve stem. (see figure 2-L) Remember that the thermostat keeps the temperature within a certain range, and usually there will be a 20 or 30 degree spread. In other words, if you set the thermostat at 350 degrees, you want the heating system to cycle *on* if the temperature is below 340 degrees, and *off* when the temperature reaches about 360 degrees. Don't forget, too, that heat rises, so the top of the oven might be warmer than thc bottom of the oven.

Get a good calibrating thermometer from your appliance parts dealer—they're pretty inexpensive. With nothing else in the oven, place it in the middle of the oven, where you can see it through the oven door glass. Let the burner cycle on and off at least twice, then observe the low temperature (when the burner cycles on) and the high temperature (when the burner cycles off.) Adjusting the screw adjusts both temperatures up or down.

2-6 (d) TIMERS AND CLOCK CONTROLS

Clocks and timers operate on 110 volt circuits, whether on gas or electric ovens. The important thing to remember about clocks in an oven unit are that often the thermostat controls are first wired through the timer. This can make them rather complex devices to troubleshoot, remove and install. Also, if you are troubleshooting a no heat complaint in an oven with an automatic cycle, the first step is *always* to check the timer controls. (An automatic cycle is one that allows you to program in advance the start and stop times and of a desired baking cycle.)

A prime example: I once got an oven service call right after Thanksgiving, where the complaint was no heat. The woman who greeted me was the ultimate Susie home-maker; everything was spotless in HER kitchen. Including the oven, which looked new, but this gal swore she'd owned it since it was new 35 years before, and by golly she knew it inside and out and didn't know why two ovens out of three suddenly didn't work. My first question: did she have family over for Thanksgiving? Yup. Does she ever use the timer function? Nope. Were her daughters or their kids using the oven? Yes again.

I reset her automatic cycle controls (which, coincidentally, were only on two of three ovens) and collected the thirty dollars for a service call, thank you very much.

Testing clocks and timer controls is just like testing switchblocks, except that the switches are opened and closed mechanically by cams inside the timer. You must have a wiring diagram to determine which terminals are connected when a switch inside the timer closes. Put a resistance meter across those terminals and advance the clock to determine if the switch is opening and closing. If the clock motor doesn't run or the switches don't open and close as they are supposed to, replace the timer. To replace, mark the wires or tag the wires with the terminal it goes on. If you need to, you can draw a picture of the terminal arrangement and wire colors. If possible, change over the timer wires one-by-one—it can be easier. If there are any special wiring changes, they will be explained in instructions that come with the new timer.

Clocks and timers can usually be rebuilt. Sometimes they have to send it away and it can take several weeks. Ask your appliance parts dealer for details.

2-6 (e) ELECTRIC HEATERS AND IGNITORS

Electric heater elements and ignitors are tested by measuring continuity across them as described in section 2-5(b). A good heater will show continuity, but quite a bit of resistance. A bad heater will usually show no continuity at all. Replace heater elements if they show no continuity.

2-6 (f) RELAYS

Heating elements use a lot of electricity compared to other electrical components. The switches that control them sometimes need to be built bigger than other kinds of switches, with more capacity to carry more electricity.

The switches involved in running a heater or ignitor can be too big to conveniently put inside the control console or timer. Besides that, there are safety considerations involved in having you touch a switch that carries that much electricity directly, with your finger.

The way they solve that problem is to make a secondary switch. A little switch activates an electromagnet, which closes a big

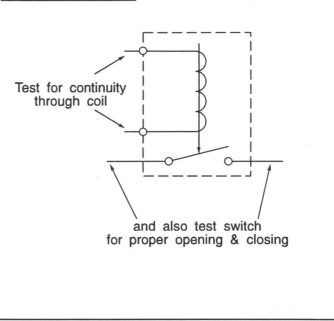

Figure 2-M Relays

Test for continuity through coil

and also test switch for proper opening & closing

switch that carries the heavier current load. This is called a relay. (Figure 2-M) To test a relay, like any other switch, you test the opening and closing of the switch, but additionally you need to test for continuity through the coil.

2-6 (g) TRANSFORMERS

Transformers are two coils wrapped around a common metal core. (Figure 2-N) Test for continuity through each coil. Also test for continuity between each lead and the metal core. This would indicate a short.

2-6 (h) SOLENOIDS

Solenoids are similar to relays; while a relay closes a switch, solenoids perform some mechanical function, like locking an oven door. They should be tested the same way. The coil should show SOME resistance, but continuity should be good. If a solenoid shows no continuity, there's a burnt or broken wire somewhere in the coil.

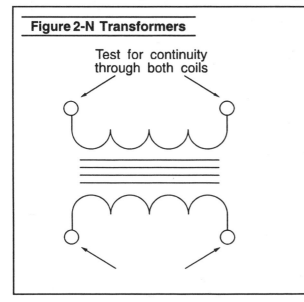

Figure 2-N Transformers

Test for continuity through both coils

Chapter 3

PARTS ACCESS

Just for your info and future reference, an *electric* cooktop and oven combined into one unit is called a **range**. A *gas* cooktop and oven combined into one unit is called a **stove**.

This manual covers both ranges and stoves, in addition to freestanding cooktops and wall ovens. The systems in freestanding cooktops and wall ovens are the same as in ranges and stoves. So when this service manual refers to a particular aspect or function of a cooktop or an oven, the same thing will be true of a range or a stove.

3-1 CONTROL PANEL

Typically clocks, timers and other manual oven controls are located in the control panel. In ranges and stoves, there may be two control panel areas. The main burner or surface unit controls and oven thermostat are usually located at the front of the cooktop area, and selector switches or lighting controls may be located in the console area, at the rear of the cooktop. In some models, the wiring diagram may be contained in an envelope inside the console. See section 3-6.

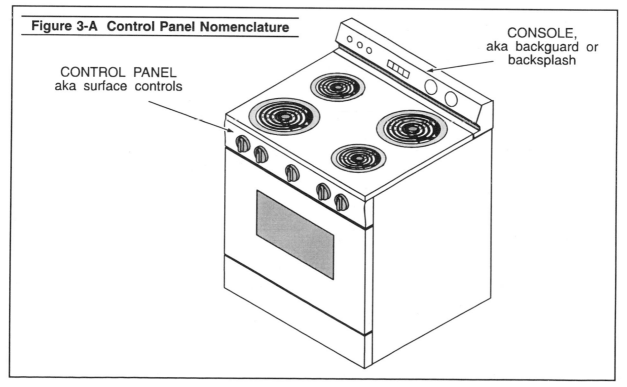

Figure 3-A Control Panel Nomenclature

CONTROL PANEL
aka surface controls

CONSOLE,
aka backguard or
backsplash

3-1 (a) UNDER COOKTOP (Figure 3-B)

In most stoves, ranges and cooktops, the cooktop just lifts up to provide access to gas inlet piping, burners and burner valves, pilots, oven thermostats, infinite switches and surface unit wiring. This, of course, makes the burners easier to clean, too. In gas models with spark ignition, the spark module may be located inside a metal box inside the cooktop compartment. Do not forget to remove power from an electric cooktop before lifting it.

3-2 BROILER PAN OR STORAGE (Figure 3-C)

Since a gas oven burner is underneath the oven, the space below it is usually dedicated to a broiler compartment. Beneath the oven, at the back of the broiler compartment is usually where the gas oven ignitor or pilot and safety valve are located. This may be different in a convection oven as shown in figure 3-D. The oven floor usually just lifts out for easier access to work on the burner, ignitor or gas valve. There will also be holes in the oven floor to facilitate airflow within the oven while it is operating.

NOTE: Do not block these holes in the oven floor with aluminum foil. It will definitely block proper airflow in your oven, and it will probably disturb burner operation, too!

Electric ovens are somewhat different. They have separate broil and bake elements at the top and bottom of the oven, respectively. So the bottom compartment is just for storage of pans, and the oven floor is not removable.

The broiler or storage pan may be removed as shown in figure 3-C (keyhole slots). In some models, the wiring diagram may be found inside the front leg of the oven after removing the broiler pan or storage drawer. See section 3-6.

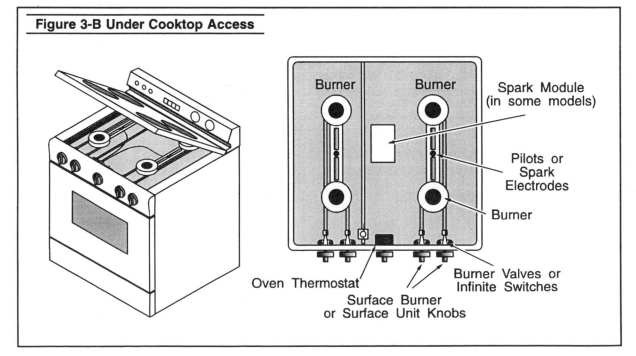

Figure 3-B Under Cooktop Access

Burner   Burner   Spark Module (in some models)

Pilots or Spark Electrodes

Burner

Oven Thermostat

Burner Valves or Infinite Switches

Surface Burner or Surface Unit Knobs

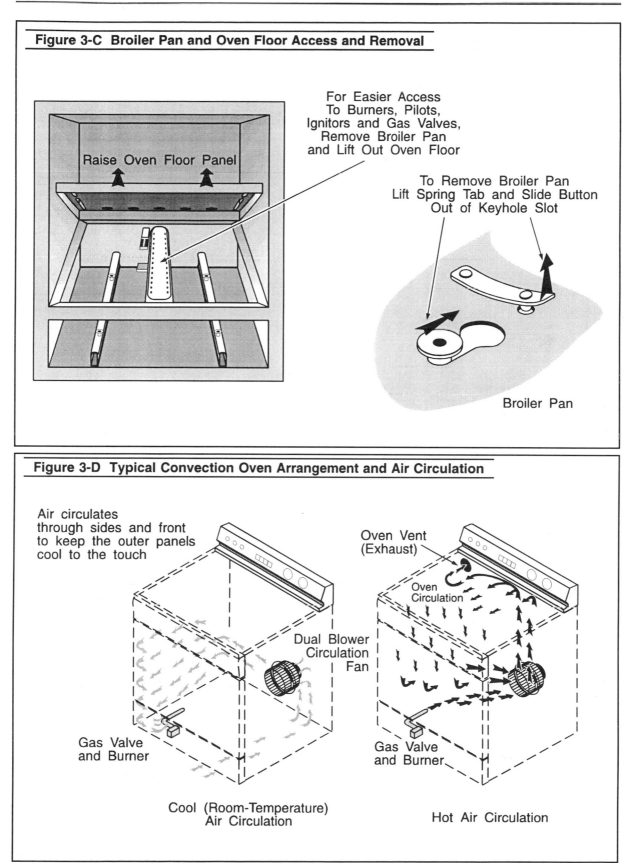

Figure 3-C Broiler Pan and Oven Floor Access and Removal

Raise Oven Floor Panel

For Easier Access
To Burners, Pilots,
Ignitors and Gas Valves,
Remove Broiler Pan
and Lift Out Oven Floor

To Remove Broiler Pan
Lift Spring Tab and Slide Button
Out of Keyhole Slot

Broiler Pan

Figure 3-D Typical Convection Oven Arrangement and Air Circulation

Air circulates
through sides and front
to keep the outer panels
cool to the touch

Oven Vent
(Exhaust)

Oven
Circulation

Dual Blower
Circulation
Fan

Gas Valve
and Burner

Gas Valve
and Burner

Cool (Room-Temperature)
Air Circulation

Hot Air Circulation

3-3 REAR ACCESS

In some models, high-limit thermostats, cleaning limit thermostats, oven thermostat probes, convection fans and sail switches, rotisserie motors, and oven spark modules are accessed by removing side or rear access panels as shown in figure 3-E. If you have a problem with burned leads to an electric oven element, sometimes you need to go in through the back to find a lost wire lead.

To access these panels, the oven obviously must first be pulled away from the wall. This can be tricky. Be careful that there is enough excess flexible gas piping and electrical power cord. If not, you must first disconnect gas and/or power from the oven.

A stove should have a flexible line from the wall valve to the the gas pressure regulator. The pressure regulator should be located under the cooktop as shown in figure 5-B. In a wall oven, you may need to open up the control panel to get to the pressure regulator.

If you can, shut off the manual gas valve at the wall. Sometimes access to it is through the kitchen cabinet adjacent to the oven or stove.

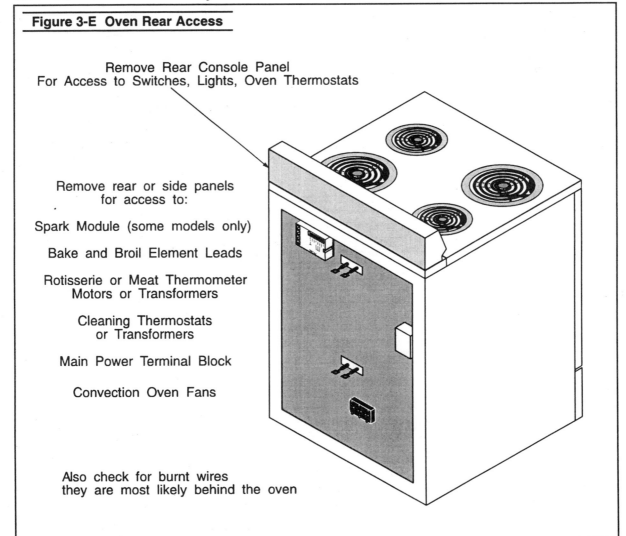

Figure 3-E Oven Rear Access

Remove Rear Console Panel
For Access to Switches, Lights, Oven Thermostats

Remove rear or side panels
for access to:

Spark Module (some models only)

Bake and Broil Element Leads

Rotisserie or Meat Thermometer
Motors or Transformers

Cleaning Thermostats
or Transformers

Main Power Terminal Block

Convection Oven Fans

Also check for burnt wires
they are most likely behind the oven

3-4 DOOR REMOVAL, DOOR SEAL AND OVEN GLASS SERVICE

In most models the door just lifts off as shown in figure 3-F. If so, open the oven door to the first notch (nearly fully closed) and lift the door straight off its' hinges. In some ovens, you must first remove one or two screws that hold the door to the hinge. Be careful and do NOT put your fingers under the hinges; the springs are strong and you might just end up wishing you hadn't.

Working on oven hinges and springs in some models requires access through the broiler/storage pan or side panels. Shown in diagram 3-F is a typical oven door hinge and spring arrangement.

Once you get the door off, remove any door edge trim and the screws that hold the door together. If you are replacing the oven glass or door seal, disassemble the door carefully and note how everything comes apart so you can get it back together. Often there are layers of stuff inside the door; insulation, oven glass (cleaning cycle) heat shields, door locking mechanisms, switches and other gadgets.

In some ovens, the door seal is attached to the oven opening instead of the door. To replace the door seal in these ovens, you sometimes have to open up the back panel of the oven and loosen the whole oven inner liner.

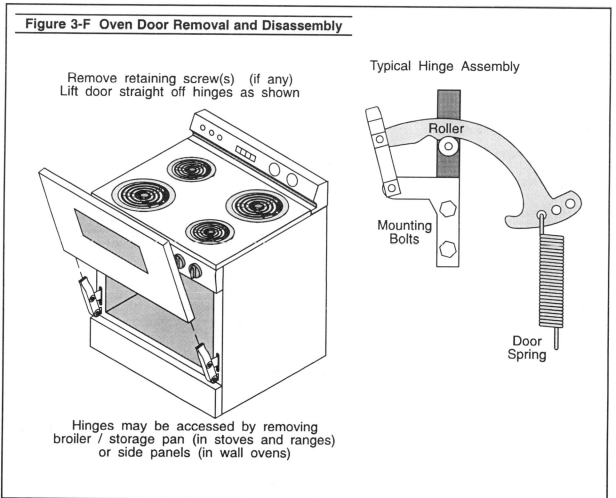

Figure 3-F Oven Door Removal and Disassembly

Remove retaining screw(s) (if any)
Lift door straight off hinges as shown

Typical Hinge Assembly

Roller

Mounting Bolts

Door Spring

Hinges may be accessed by removing broiler / storage pan (in stoves and ranges) or side panels (in wall ovens)

3-5 NAMEPLATE INFO

The metal nameplate is usually found inside the door or under the cooktop as shown in figure 3-G. It may also be fastened to the top edge of the door itself, on top of the console, or inside of the broiler compartment or pan storage compartment.

If you cannot find the nameplate, check the original papers that came with your oven when it was new. They should contain the model number somewhere.

In any case, and especially if you have absolutely NO information about your oven anywhere, make sure you bring your old part to the parts store with you. Sometimes they can match it up by looks or by part number.

3-6 WIRING DIAGRAM

In older models, wiring diagrams may be pasted to the back of the oven. In newer machines, they may be in a plastic bag inside the console or inside one of the front legs of the oven. Access as shown in figure 3-G.

If you are tracing a complex electrical circuit, it may not hurt to make a couple of photocopies of the wiring diagram, so that you can physically follow different circuits with a colored highlighter pen.

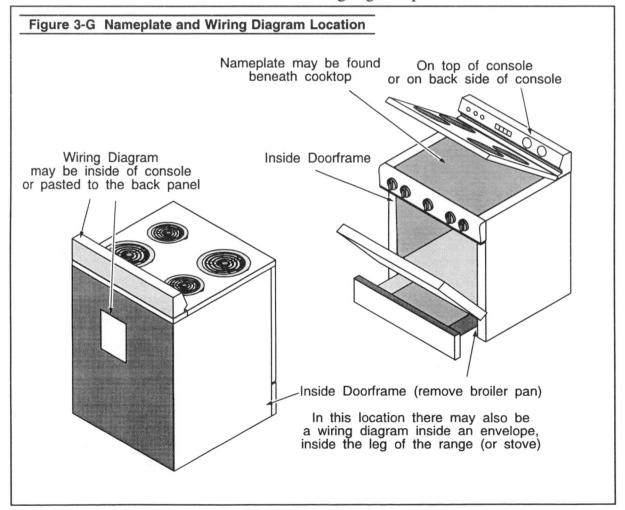

Figure 3-G Nameplate and Wiring Diagram Location

Nameplate may be found beneath cooktop

On top of console or on back side of console

Wiring Diagram may be inside of console or pasted to the back panel

Inside Doorframe

Inside Doorframe (remove broiler pan)

In this location there may also be a wiring diagram inside an envelope, inside the leg of the range (or stove)

Chapter 4

ELECTRIC COOKTOPS AND OVENS

4-1 NORMAL OPERATION

NOTE: The diagnosis sections of this chapter assume that all other electrical controls, i.e. timer and thermostatic-limit controls, as described in chapter 2, are operating properly, and the malfunction has been isolated to the heating system!!!

The heating element is simply a big resistor wire, with enough resistance to generate a high heat. Usually these are nichrome wire, surrounded in ceramic insulation, with a steel sheath around the ceramic.

Inside the oven, the heating element is called a "bake" or "broil" element. A "bake" element is located below the oven it affects. A "broil" element is located above the oven it affects. See figure 4-A.

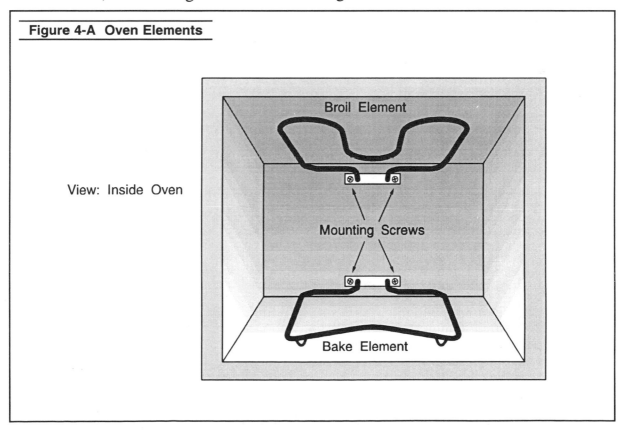

Figure 4-A Oven Elements

View: Inside Oven

Broil Element

Mounting Screws

Bake Element

In a cooktop, the heater elements are coiled into a round thing called a "surface unit." A single surface unit might contain two or even three different elements all mounted together, with different resistance ratings. Sometimes the wiring is screwed directly to terminals on the ends of the surface unit. Some surface units plug into a receptacle mounted under the cooktop, which makes them more easily removable for cleaning, but also more susceptible to burned connections. (Figure 4-B)

TEMPERATURE CONTROL

To maintain a set temperature in a *cooktop*, the element is cycled on and off, usually by a switch called an infinite switch, so named because it theoretically provides an infinite number of heat settings. This switch has its own little heater inside, which heats a bimetal switch. (see figure 4-C) A cam attached to the control knob changes spring tension on the bimetal, which changes the amount of heat needed to open the switch. When the heating

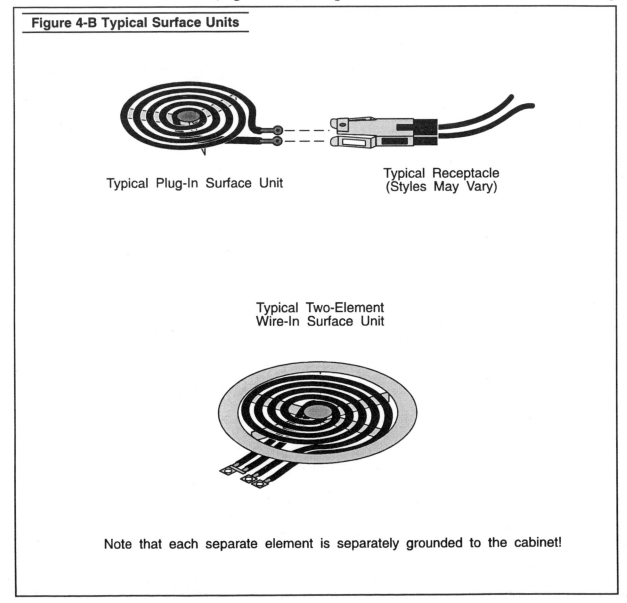

Figure 4-B Typical Surface Units

Typical Plug-In Surface Unit

Typical Receptacle
(Styles May Vary)

Typical Two-Element
Wire-In Surface Unit

Note that each separate element is separately grounded to the cabinet!

element is on, the heater inside the switch is on. The bimetal heats (along with the element) until the contacts open. Then the bimetal cools (along with the elements) until the contacts close again.

There are also fixed-temperature switches that vary the voltage going to the heating elements to maintain fixed, pre-set temperatures. These are usually push-button or rotary switches with fixed settings such as warm, low, me-dium and high. In fixed-temperature switch controls, heat levels are varied by applying different voltages (110V or 220V) to different coils of different re-sistances, as shown in figure 4-C.

In an *oven*, The temperature is con-trolled by a thermostat. Using a liquid-filled bulb and capillary, the thermostat senses temperature inside the oven and cycles the heating system on and off to maintain oven temperature within a certain range.

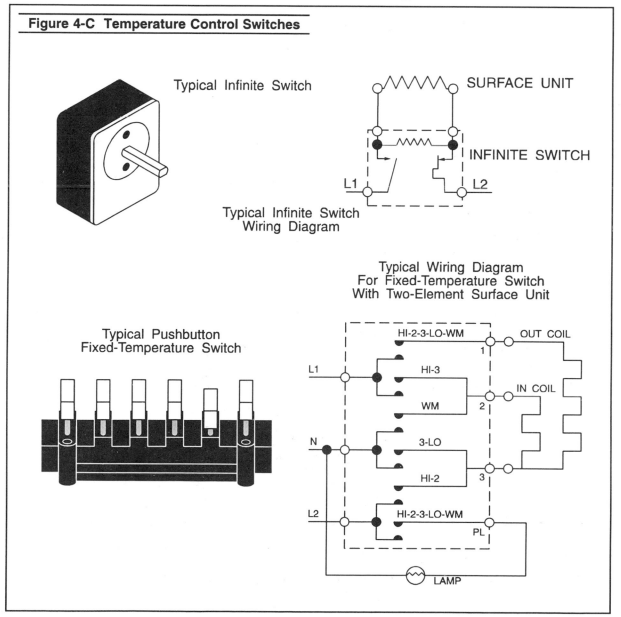

Figure 4-C Temperature Control Switches

Typical Infinite Switch

SURFACE UNIT

INFINITE SWITCH

Typical Infinite Switch Wiring Diagram

Typical Pushbutton Fixed-Temperature Switch

Typical Wiring Diagram For Fixed-Temperature Switch With Two-Element Surface Unit

LAMP

4-2 CLEANING HEATING ELEMENTS

Do not put any cleaning agents or solvents directly on heater elements or surface units. The steel sheath is semi-porous; cleaning solvents can penetrate the steel and damage the ceramic insulation or electrically short the element. Think about it... when these elements are in operation, they glow red-hot. They will eventually incinerate anything that contacts them. If there is any carbon or other crusty residue left after that, just scrape it off as best you can.

The exception to this is the solid, or "euro-style" surface units. These units can be cleaned; your appliance parts dealer has a special cleaner for this purpose.

4-3 TROUBLESHOOTING

When troubleshooting electrical cooking equipment, the very first thing to do is stand back and observe the big picture. What's really going on? If *nothing* is operating, you probably have a breaker, fuse or other power source problem. Do the surface units heat, but not the oven elements? Does the bake element heat, but not the broil element? Does the oven cleaning only work on Tuesdays in July during a snowstorm? Knowing what's operating and what isn't, in conjunction with a wiring diagram, can point you towards the failed component.

In a moment, we'll talk about the general steps to follow. But first, I want to impress upon you something really important. I know I said this in chapter 2, but it bears repeating. In electric cooking equipment, you're usually dealing with 220 volt circuits. DO NOT TAKE THIS LIGHTLY. *220 VOLTS CAN KNOCK YOU OFF YOUR FEET, AND DO YOUR BODY SERIOUS DAMAGE, VERY QUICKLY. DO NOT TEST LIVE 220 VOLT CIRCUITS.*

Unless you are dealing with an obvious, simple repair like a burned element, isolating the problem in a 220 volt cooking circuit basically boils down to shutting off the power, testing each component for continuity, and looking for burned or chafed wires. Electric oven or cooktop repairs can be broken down into 3 categories:

4-3 (a) ONE ELEMENT NOT HEATING

In surface units, this is usually caused by a burned out element, terminal, or receptacle. (see figures 4-B and 4-D) It may also be an infinite switch or fixed-temperature selector switch. Turn the breaker off or pull the fuse, and inspect the element, terminals and receptacle.

Most receptacles are mounted to the cooktop by one or two screws. Some receptacles can be disassembled as shown in figure 4-D to inspect and replace the internal terminals.

Usually the burned or melted area of terminals or elements will be visible, but test for continuity, even if it appears to be OK. A bad element will show no continuity. A good element will show some continuity, even though there is a lot of resistance.

A switch will show no continuity when off and good continuity when turned on. You should also see continuity through the bimetal heater inside an infinite switch.

Replace a burned or bad receptacle, terminal, element or switch; repair a broken wire end terminal. When replacing elements, make sure you get a replacement element of the right wattage; the element is matched to the control switch.

If an *oven* element isn't working, do not forget rule number one from section 2-6 (d): check the automatic baking cycle (timer) controls first! If those are OK, the break is usually where the wire attaches to the element, inside the back wall of the oven. Turn the breaker off or pull the fuse, remove the screws holding the element into the oven, and pull the element away from the back wall a little. There is a little bit of extra wire in there to allow you to access it from the front, but do not pull out any more wire than you need to work on it. You may need to tilt the element upwards to get the terminals through the holes. If the wire is broken or burnt completely off the terminal, you may be able to fish the wire out of the hole with needlenose pliers, as long as the power is off.

If there is a burn in the element, usually it will be visible, but test the element for continuity, even if it appears to be OK. A bad element will show no continuity. A good element will show some continuity as described in section 2-6(e).

Replace a burned or bad element; repair a broken wire end terminal using special high-temp terminals, available at your appliance parts dealer.

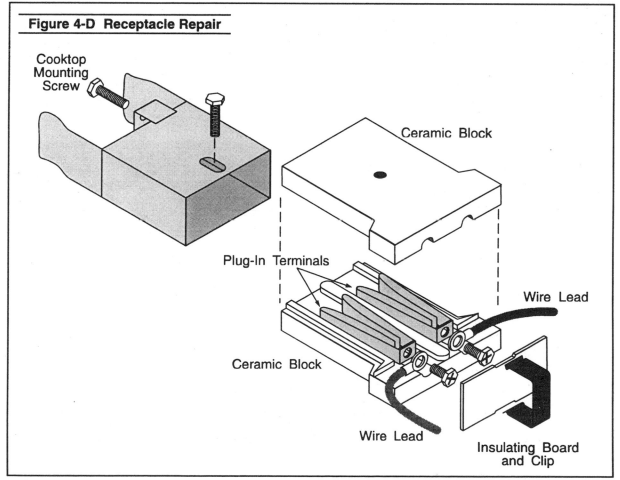

Figure 4-D Receptacle Repair

Cooktop Mounting Screw

Ceramic Block

Plug-In Terminals

Ceramic Block

Wire Lead

Wire Lead

Insulating Board and Clip

4-3(b) NO POWER TO OVEN

First, make absolutely sure you have no power. Check all heating elements, on all settings. Check the clock, timers, and the oven light if any. If at least one component is operating properly, you have power. Keep in mind that if one leg of the circuit is out due to a house wiring problem, you might have 110 volts but not 220. (see section 2-5, page 16) This might mean the oven light and clock and even some of the heating elements operate on low power, but the high-heat circuits do not work. It also means you still have power, and you can still get zapped.

If you're sure you have no power, we need to figure out if the problem is in the house wiring or inside the oven or cooktop. First, of course, check the house breaker or fuse. Next we need to test for power where it enters the oven.

This can sometimes present a problem. In most installations, there is a 220 volt wall plug. If this is the case, turn the breaker off or pull the fuse, pull the oven or range away from the wall, pull the plug out of the wall, turn the breaker back on, and test the wall outlet as described in chapter 2. Also check the terminal block for problems as described below. The terminal block is where the main power cord attaches to the oven circuitry. It will be just inside the back somewhere. (see figure 4-E)

In some installations, the oven or cooktop is wired directly into the house wiring. If so, the wiring will be connected directly to a terminal block within the unit. You need to follow the steps as described above, but while the power is off, locate the 3-wire terminal block as shown in figure 4-E.

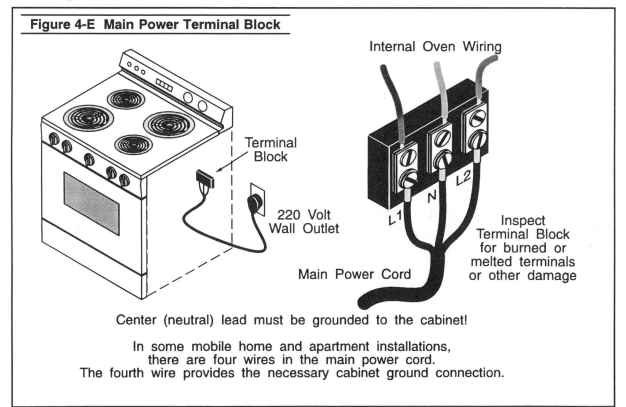

Figure 4-E Main Power Terminal Block

Terminal Block

220 Volt Wall Outlet

Internal Oven Wiring

L1   N   L2

Main Power Cord

Inspect Terminal Block for burned or melted terminals or other damage

Center (neutral) lead must be grounded to the cabinet!

In some mobile home and apartment installations, there are four wires in the main power cord. The fourth wire provides the necessary cabinet ground connection.

Inspect the terminal block for any signs of damage; overheating, melted terminals, etc.

Make sure all wiring is clear and make sure you don't touch any bare wires or terminals, turn the breaker or fuse back on briefly, and check the terminal block for power across all three legs as shown in chapter 2. Then remove power again at the breaker or fuse.

If power is not getting to the terminal block, the problem is in your house wiring. During the 70's some houses were built with aluminum wiring, which is notorious for not being able to handle oven currents. House wiring repairs are beyond the scope of this manual. There are plenty of good books on house wiring; get one of those, or call an electrician.

If power is getting to the terminal block, the problem is obviously somewhere within the oven. There may be a main fuse, or a main switch that everything is routed through. Find your wiring diagram, isolate the problem and repair as described in section 4-3(c).

4-3 (c) COMPONENT PROBLEMS

One fairly common failure with confusing symptoms occurs when an infinite switch or a fixed-temperature switch shorts internally to ground. The symptom will be that with the switch off, the indicator light remains on dimly. If this occurs, replace the defective switch.

Let's say that you have power to the surface units but none to either the bake or broil element. Or let's say the oven comes on, but the self-clean function doesn't work. It's time for a wiring diagram. Find the wiring diagram for your machine as described in chapter 3. Trace the circuits as best you can and test components as described in chapter 2. The wiring diagram can be very complex and difficult to follow, but the general objective is to trace the malfunctioning circuit as described in section 2-5, find the components in it and check them for continuity. Also check for burned wires, especially in the back of the oven. It might help to make several copies of the wiring diagram and trace the circuit that you're interested in with a colored highlighter pen.

Replace the bad component, or repair damaged wires with special high-temp wire and connections, available at your appliance parts dealer.

Chapter 5

GAS COOKTOPS

5-1 NORMAL OPERATION

Temperature is controlled in a surface burner by varying the flow of gas to it with a manual gas valve. Ignition is achieved in one of two ways; either a standing pilot flame or a spark ignition. Gas shoots out of jets on the side of the burner and into a flame tube, which carries the gas to the ignition source. The flame then chases the gas back up the tube to ignite the burner. (See figure 5-A) This allows two or more burners to share the same ignition source.

TROUBLESHOOTING

The most common problems with gas burners is that a pot boils over and stuff clogs the gas flame tube jets. In spark ignition systems, another common problem is that the ignitor switch fails and the ignitor doesn't spark for one burner, or the switch shorts and the electrodes won't stop sparking at all.

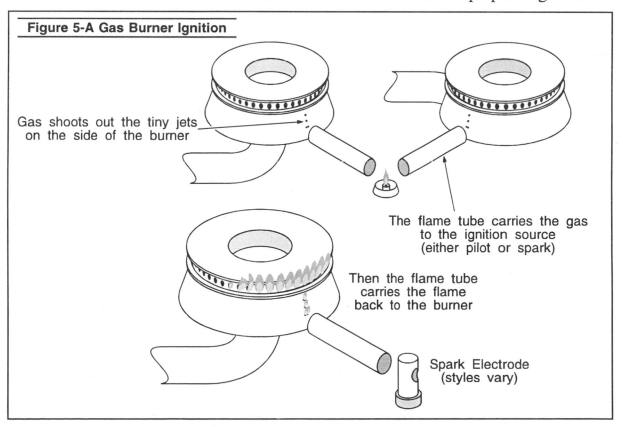

Figure 5-A Gas Burner Ignition

Gas shoots out the tiny jets on the side of the burner

The flame tube carries the gas to the ignition source (either pilot or spark)

Then the flame tube carries the flame back to the burner

Spark Electrode (styles vary)

5-2 PILOT IGNITION

Gas is supplied to the pilots by tubes coming directly off the main gas header (manifold.) There will usually also be a filter. (see figure 5-B) Somewhere in the gas line to the pilot there will be at least one pilot adjustment screw, usually on the filter or at the pilot itself. The pilots rarely need adjustment. If you do need to adjust them, they usually need to be set at about 1/4 inch tall, mostly blue flame with just a tinge of yellow at the tip. If there is a shield around them, you do not want the flame touching the shield. If turning on the burner makes the pilot blow out, either it is partly clogged with ash (See section 5-4) or the flame is too low.

5-3 SPARK IGNITION

The advantage of spark ignition systems is that they do not need a constant pilot, and thus do not waste as much gas as pilot systems. The burners use a flame tube system as described above, except that in place of the pilots, there are electrodes that spark to ignite the gas.

Spark ignition systems use a spark module to generate a pulsing, high-voltage spark to ignite the gas. (figure 5-C) The spark module is an electronic device that produces 2-4 high-voltage electrical pulses per second. These pulses are at very low amperage, measured in milliamps, so the risk of shock is virtually nil. But the voltage is high

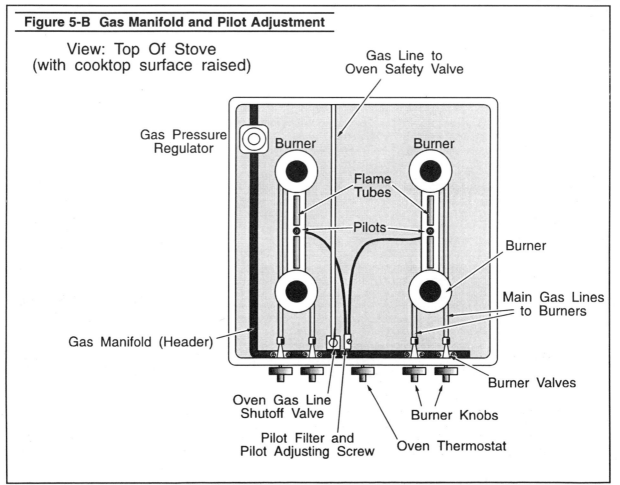

Figure 5-B Gas Manifold and Pilot Adjustment

View: Top Of Stove
(with cooktop surface raised)

Gas Line to
Oven Safety Valve

Gas Pressure
Regulator

Burner

Burner

Flame
Tubes

Pilots

Burner

Main Gas Lines
to Burners

Gas Manifold (Header)

Burner Valves

Oven Gas Line
Shutoff Valve

Burner Knobs

Pilot Filter and
Pilot Adjusting Screw

Oven Thermostat

enough to jump an air gap and ig-nite gas. When the spark is operating, you hear a tick-tick-tick, 2-4 ticks per second.

The spark ignition module (figure 5-C) is usually located inside a metal box under the cooktop surface or inside the back panel of the stove. To find the spark module on your cooktop, follow the electrical leads from the spark electrodes; they will attach directly to the module.

A rotary switch attached to the gas valve spindle closes to activate the spark module. (see figure 5-D) All cooktop ignitors spark at once; in some stoves, the oven ignitor sparks, too.

TROUBLESHOOTING SPARK IGNITION

Trouble with surface spark ignition systems usually shows up in one of two ways. Either the electrode will not spark, or it will not *stop* sparking.

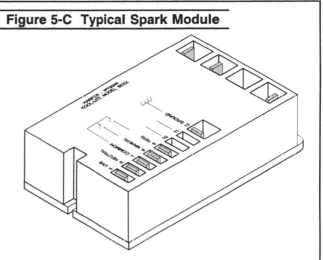

Figure 5-C Typical Spark Module

When troubleshooting cooktop electrodes keep something in mind. In some cooktops, the electrodes are wired in series; in others, they are wired in parallel. If the electrodes are in series, and one is not sparking, none will spark.

Also, all spark electrodes use the same spark module, but different switches. So if one burner is not igniting, while the gas valve is still open to that burner, try turning on another burner that doesn't share the same ignitor. In this way you can isolate the problem to the switch (which it is, 99 percent of the time,) the electrode, or the spark module.

To confirm your diagnosis, pull the leads from the switch, touch them together and see if the ignitor sparks. In testing these switches, do not forget that they operate on 110 volts. If you get too fast and loose with pulling the leads off to test them, you might zap yourself. Try using two pairs of insulated needlenose pliers to pull the leads off the switch.

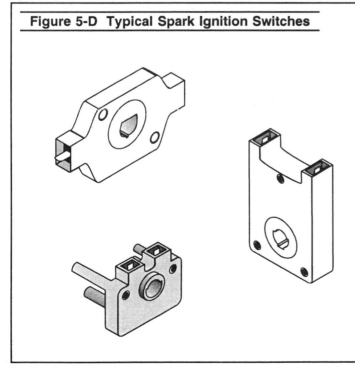

Figure 5-D Typical Spark Ignition Switches

If the ignitors will not stop sparking, usually one of the ignitor switches has shorted, due to moisture getting into it. To figure out which, pull one lead off each switch, one at a time, until the ticking stops. Replace the defective switch. In certain cooktops, a failure in the spark module will cause this symptom. If none of the switches seem to be defective, replace the spark module.

5-4 PILOTS AND BURNER ORIFICE MAINTENANCE & REPAIR

Natural gas, they tell us, is clean burning. Well, it makes good ad copy, but it's not 100 percent true. There are trace impurities in gas, and when they burn they become ash. And over a long long period of time this ash can build up and clog tiny gas orifices, like pilot orifices. The symptoms may be that the pilot will not stay lit, or blows out too easily.

There is also a little "cup" around the pilot light in a surface burner to regulate air for proper combustion of the pilot flame. These can get covered with ash or clogged up with spillage from the burner (see below.) Make sure you clean the pilot area thoroughly.

You can usually clean them out with an old toothbrush and some compressed air, but pilot orifices are generally so inexpensive that it's cheaper and safer to replace them. If you choose to clean them out, use a soft-bristle brush like a toothbrush, and not a wire brush. A wire brush might damage the orifice. Be careful not to push the ash into the orifice and impact it.

The surface burner jets are a different animal. The problem is, that pot that boils over can leave some pretty crusty stuff on the burners. The challenge is to get that crusty stuff off without enlarging the gas holes. Sometimes the crusty stuff is not crusty but gummy, and this can be even more fun to get off the burner.

There's no magic way to do this. Depending on what's clogging the gas holes, you may be able burn it off, or at least char it so it can be scraped or knocked off. Use a propane torch or one of the other stove burners if you can get it into the correct position.

You may also be able to use alcohol or some other non-petroleum solvent to knock it loose. If you do, make sure the solvent is completely gone before you put the burner back into service. (Petroleum-based solvents might leave a residue)

Try poking it out with a straight pin if possible. You can use a welding tip cleaner, or a small twist drill, to poke into the holes but only if it fits in the hole with plenty of clearance. Do not use a drill motor with a twist drill; just push it in and out by hand. Use it as if it was a file. Use a twist drill or tip cleaner that is MUCH smaller than the diameter of the hole. Do not use anything with a remote chance of enlarging the hole. Remember, the objective is to take off the crusty stuff without taking off any of the metal. Sometimes a burner is just too badly clogged, or the holes are too small, and the burner must be replaced.

Chapter 6

GAS OVENS

6-1 NORMAL OPERATION

NOTE: The diagnosis sections of this chapter assume that all other electrical controls, i.e. timer and thermostatic controls, as described in chapter 2, are operating properly, and the malfunction has been isolated to the ignition or heating system!!!

Gas ovens are quite different from gas cooktops; since the burner is inside the oven, you cannot immediately see whether the burner has ignited. If it does not ignite, you certainly do not want the gas valve to stay open. This would dump raw unburned gas into the oven and create an explosion hazard. To avoid this, designers use a gas *safety valve* that does not open until ignition is assured.

Different manufacturers have designed different methods of keeping the gas safety valve closed until ignition is guaranteed. Some safety valves use low voltage, some high voltage, and some use hydraulic pressure. It is important to know which you are dealing with, because many of the valves look the same. If you try to test a low voltage valve by putting high voltage across it, you will burn it out.

Do not confuse the oven gas thermostat valve with the safety valve. The oven gas *thermostat valve* is a valve you set by hand (the oven temperature knob) to control the oven temperature. In some types of systems, the thermostat is not a valve at all; it is an electrical switch that opens or closes based on the oven temperature it senses. The gas *safety valve*, on the other hand, simply prevents gas from flowing to the burner until ignition is guaranteed. In systems with an electrical thermostat, the safety valve opens and closes to cycle the burner on and off, but it still will not open if there is no ignition.

The nice thing about gas ovens is that there aren't a whole lot of moving parts, so wear and tear is a relatively minor consideration. In diagnosing them, you basically just have to learn how the system is *supposed* to work, then watch it for a while to see which part of the system has *stopped* working.

The different systems are detailed in the following sections.

6-2 PILOT IGNITION

Pilot ignition systems use a flame sensing element to sense whether the pilot is lit and the safety valve can open. The sensing element sits right in the pilot flame.

Just exactly *where* the sensor sits in the pilot flame is important. (See figure 6-A) If the sensing bulb is not in the right part of the flame, or if the pilot is adjusted too low or too high, it will not get hot enough and the safety valve will not open.

In different systems, the sensor uses one of two methods to open the safety valve: capillary systems or millivolt systems.

6-2 (a) MILLIVOLT PILOT SYSTEMS

When two dissimilar metals (for example, copper and steel) are bonded together electrically, and then heated, they generate a tiny electrical current between them. The voltage is very small, measured in millivolts. This is the basis for a millivolt oven ignitor system. All that's needed is a safety valve that will sense this tiny voltage and open the valve if it is present. If the pilot is out, there is no millivoltage and the safety valve will not open. See figure 6-B.

The bimetal that generates the millivoltage is called a pilot generator. It generates about 750 millivolts, or about 3/4 volt.

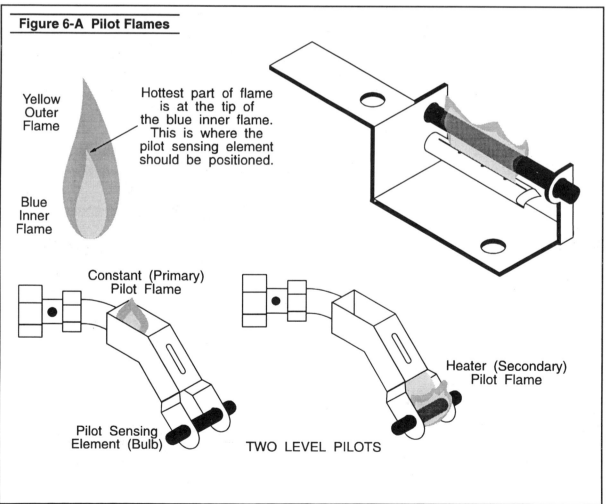

Figure 6-A Pilot Flames

Yellow Outer Flame

Hottest part of flame is at the tip of the blue inner flame. This is where the pilot sensing element should be positioned.

Blue Inner Flame

Constant (Primary) Pilot Flame

Pilot Sensing Element (Bulb)

Heater (Secondary) Pilot Flame

TWO LEVEL PILOTS

TROUBLESHOOTING AND RE-PAIR OF MILLIVOLT SYSTEMS

First, of course, if the oven has an automatic baking cycle as described in section 2-6(d), check the automatic cycle (timer) controls.

If the burner in a millivolt system will not start, typically the problem is the gas valve. Occasionally the problem might be the pilot generator or thermostat. The thermostat in these is just a temperature-sensitive on/off switch. To test, turn it on and test for continuity. Try cleaning the pilot orifice and pilot generator as described in section 6-5 and adjusting it a little higher. The pilot generator needs to be sitting right in the hottest part of the flame as described in section 6-2.

If that doesn't work, we have a minor dilemma in determining whether the problem is the pilot generator or the safety valve. The dilemma here is that the voltages are too small to be measured with standard equipment. VOM millivolt adaptors cost nearly as much as the pilot generator itself. And the safety valve, which is usually the problem, costs twice as much as the pilot generator. So usually you just replace either or both of them. But don't forget they are electrical parts, which are non-returnable. What I recommend is just to replace the gas valve first; that usually will solve the problem. If not, replace the pilot generator. You just ate a gas valve, but trust me, you'd have bought one sooner or later anyway.

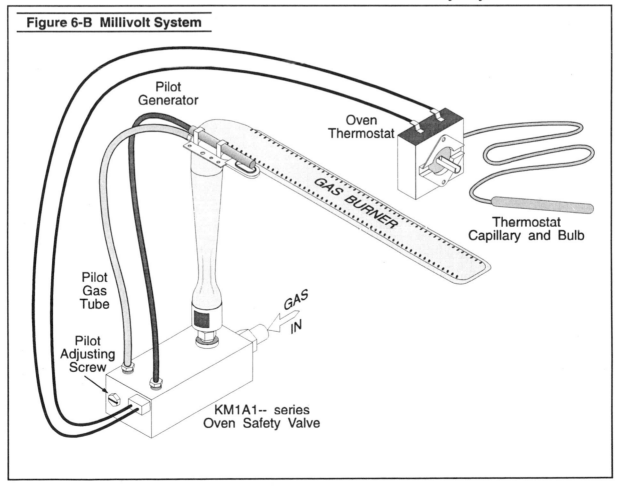

Figure 6-B Millivolt System

When installing the pilot generator, screw it into the safety valve finger tight, plus 1/4 turn. Any tighter than that and you can damage the electrical contacts on the valve.

6-2 (b) CAPILLARY PILOT SYSTEMS

In some systems the sensor is a liquid-filled bulb, with a capillary to the safety valve or flame switch. When the liquid inside heats up, it expands and exerts pressure on a diaphragm, which opens the valve or closes the switch.

It is important to know that these sensor bulbs *do not* cycle the burner on and off to maintain oven temperature. That is the thermostat's function. It has a sensor bulb too, but it senses oven temperature, not pilot flame. The only function of these pilot sensing elements is to prevent gas flow to the burner if the bulb does not

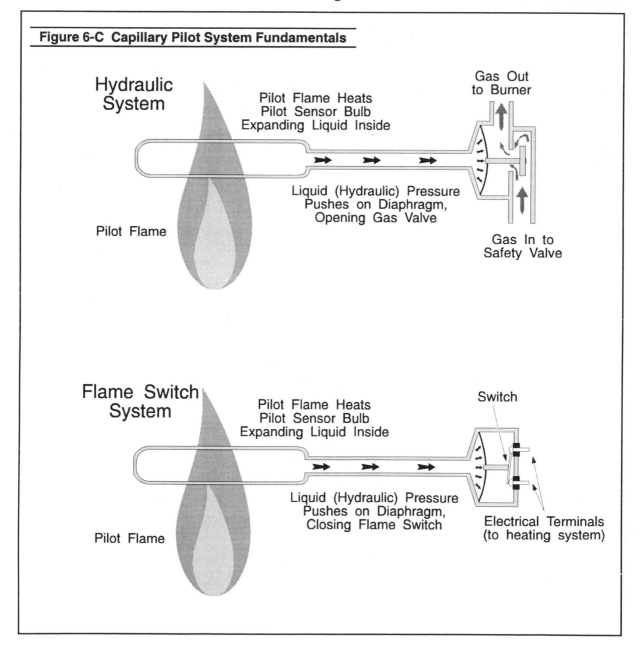

Figure 6-C Capillary Pilot System Fundamentals

get hot enough to assure burner ignition.

In *flame switch systems*, hydraulic pressure from the capillary physically closes the switch, which completes an electrical circuit to the safety valve. The safety valve is electrical and operates on 110 volts. See Figure 6-D. If the pilot is out, the flame switch does not close and the 110 volt heating circuit is not complete, so the safety valve will not open.

In *hydraulic capillary systems*, hydraulic pressure from the capillary physically opens the gas safety valve. Figure 6-E shows one type of these.

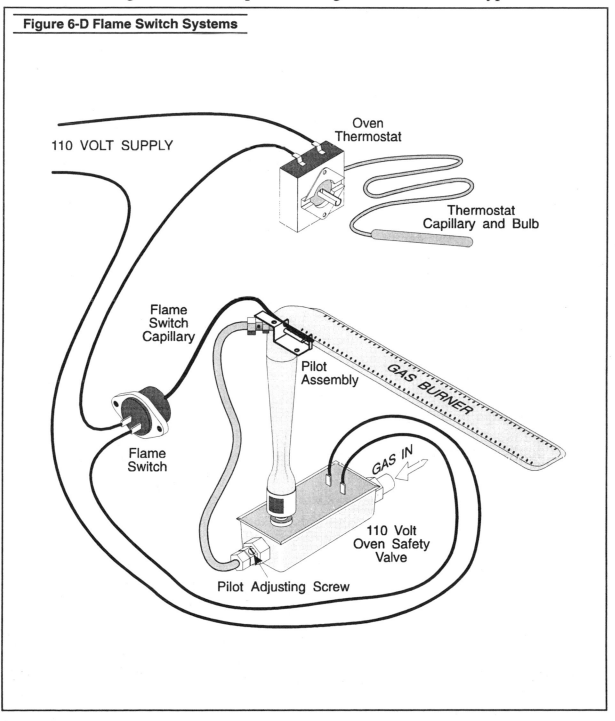

Figure 6-D Flame Switch Systems

TWO-LEVEL PILOTS

Some of these direct-pressure (hydraulic) systems use a two-level pilot. The pilot stays at a very low level; not even high enough to activate the safety valve. This is called the *constant* pilot, or *primary* pilot. Gas for the primary pilot may come from either the thermostat or directly from the gas manifold.

When the thermostat valve is turned on, the pilot flame gets bigger, heating the sensor bulb, which activates the safety valve (hydraulically) and the burner ignites. This is called the *heater* pilot, or *secondary* pilot. Gas for the secondary pilot comes from the oven thermostat itself.

When the oven reaches the correct temperature setting, the thermostat drops the pilot flame back to the lower level, the safety valve closes and the burner shuts off. See figure 6-E.

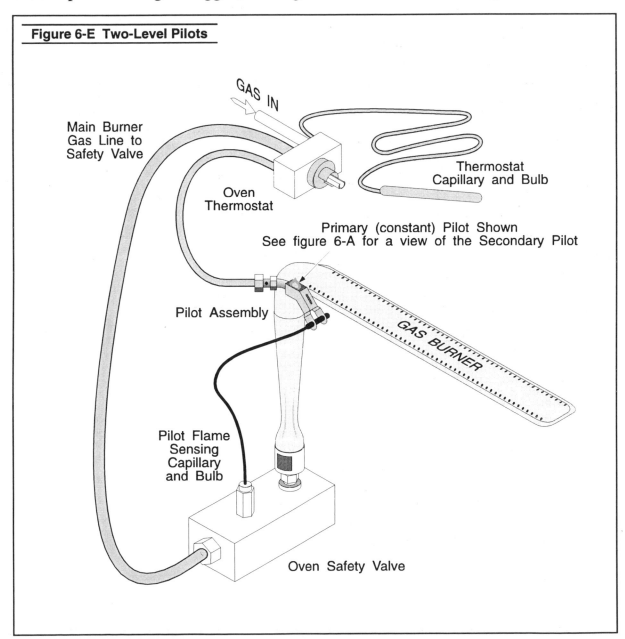

Figure 6-E Two-Level Pilots

GAS IN

Main Burner
Gas Line to
Safety Valve

Thermostat
Capillary and Bulb

Oven
Thermostat

Primary (constant) Pilot Shown
See figure 6-A for a view of the Secondary Pilot

Pilot Assembly

GAS BURNER

Pilot Flame
Sensing
Capillary
and Bulb

Oven Safety Valve

TROUBLESHOOTING & REPAIR OF CAPILLARY PILOT SYSTEMS

If the oven burner won't ignite, as always, first check the automatic bake cycle (timer) controls, if any, as mentioned in section 2-6(d).

The sensing bulb needs to be sitting right in the hottest part of the flame as described in section 6-2. If you don't have a good strong pilot (secondary pilot, in two-level systems) that engulfs the pilot sensing bulb with flame, try cleaning the pilot assembly and sensor bulb as described in section 6-5. If that doesn't work, replace the pilot assembly.

If you *do* have a good strong pilot that engulfs the pilot sensing bulb with flame, then odds are that the sensing element and/or whatever it is attached to are defective. If it is a flame switch, replace the flame switch. If it is a safety valve replace that.

In a two-level pilot system, remember that the main oven thermostat supplies the secondary pilot with gas. So if you cannot get a good secondary pilot the problem may be the pilot assembly, *or it may be the thermostat.* If you do get a good secondary pilot, you're back to the sensing bulb and safety valve.

Replace the defective component.

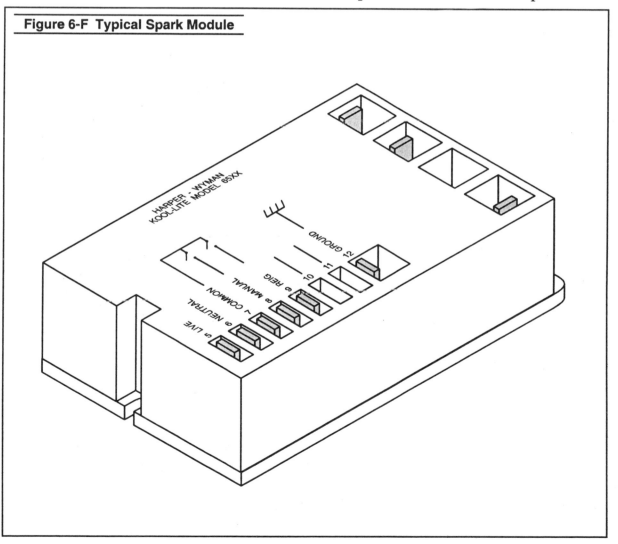

Figure 6-F Typical Spark Module

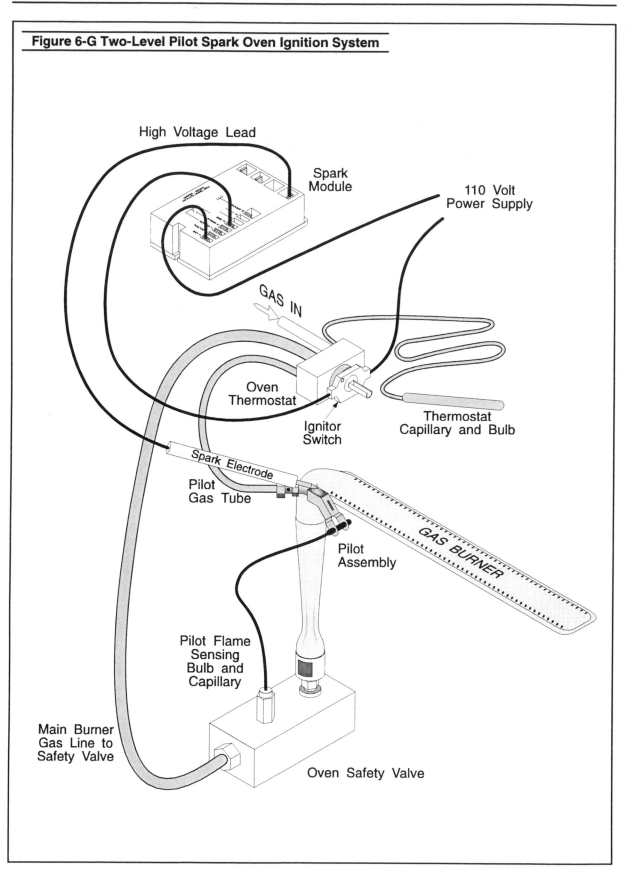

Figure 6-G Two-Level Pilot Spark Oven Ignition System

6-3 SPARK IGNITION

The advantage of spark ignition systems is that they do not need a constant pilot, and thus do not waste as much gas as pilot systems.

Spark ignition systems use a spark module to generate a pulsing, high-voltage spark to ignite the gas. The spark module is an electronic device that produces 2-4 high-voltage electrical pulses per second. These pulses are at very low amperage, measured in milliamps, so the risk of shock is virtually nil. But the voltage is high enough to jump an air gap and ignite gas. The spark ignition module is usually located either under the cooktop or inside the back of the stove. The same module is used for both the surface burner ignition and the oven burner ignition.

However, the spark is not certain enough to light the oven burner, and the gas flow is too high, to rely on the spark alone. Remember, in an oven, before the safety valve opens, you need to be **assured** of ignition. So the spark ignites a low-gasflow pilot, and then the safety valve opens only when the *pilot* is lit.

TWO-LEVEL PILOT
SPARK SYSTEM (figure 6-G)

This is the same two-level pilot system described in section 6-2(b), with a few important exceptions. The *constant* or *primary* pilot does not stay lit when the oven thermostat is turned off. It does, however, stay lit the whole time the oven thermostat is turned on.

When the oven is turned on, a switch mounted to the oven thermostat stem signals the spark module. These are the same switches as shown in section 5-3.

The flame is positioned between the spark electrode and its target. The pilot flame actually conducts electricity. So when the pilot flame is burning, electricity from the spark electrode is drained off to ground, and sparking stops. If the pilot quits, sparking resumes.

When the thermostat calls for more heat in the oven, the *heater* or *secondary* pilot increases the size of the pilot flame, which heats the sensing bulb, which opens the safety valve and kicks on the burner.

When oven temperature is reached, the thermostat drops the pilot back to the lower level, the pilot sensing bulb cools and the burner shuts down.

PILOT VALVE SPARK SYSTEM

Yup, this ol' boy's got it all. Spark ignition, a pilot, a flame switch and TWO—count 'em—TWO safety valves; one for the pilot and one for the burner. (Figure 6-H)

The operation is actually simpler than the diagram looks. When you turn on the oven thermostat, a cam on the thermostat hub closes the pilot valve switch. This opens the 110 volt pilot safety valve and energizes the spark module, igniting the pilot. As in the other spark system, the pilot flame provides a path that drains off the spark current, so the ignitor stops sparking while the pilot is lit. As long as the oven thermostat is turned on, the pilot valve switch stays closed, so pilot valve stays open and the pilot stays lit.

When the pilot heats the pilot sensing element of the flame switch, the flame switch closes. This completes the 110 volt circuit to the oven safety valve, so the valve opens and the burner ignites.

When the oven temperature reaches the set point of the thermostat, the thermostat switch opens, breaking the circuit and closing the oven safety valve, and shutting off the burner.

TROUBLESHOOTING AND REPAIR OF SPARK OVEN IGNITION SYSTEMS

If the oven won't ignite, as always, first check the automatic bake cycle (timer) controls, if any, as mentioned in section 2-6(d).

Now that you know how the system works, first look to see what is not working. When the oven thermostat is on, and there isn't a pilot flame, is the electrode sparking? Is there spark, but no primary pilot? Is the primary pilot igniting, but not the secondary? Is there sparking after the thermostat is shut off?

IF SPARKING OCCURS

(The pilot may or may not light, but the main burner is not lighting) Remember that the thermostat supplies the pilot with gas in these ovens, and only when the thermostat is on. So if you don't have a primary *and* secondary pilot flame, odds are the problem is the pilot orifice or oven thermostat. Try cleaning the pilot assembly and sensor bulb as described in section 6-5. If that doesn't work, adjust the secondary flame a little higher. If that doesn't work, replace the pilot assembly.

If you *do* have a good strong secondary pilot that engulfs the pilot sensing bulb with flame, then odds are that the oven safety valve (or flame switch, whichever is attached to the pilot sensing bulb in your system) is defective. Replace the defective component.

IF SPARKING DOES NOT OCCUR

Something is wrong with the high-voltage sparking system. If you are in a hurry to use your oven, you can turn on the oven thermostat, carefully ignite the primary pilot with a match and use the oven for now; but remember that the minute you turn off the thermostat, the pilot goes out.

Are the cooktop ignitors sparking? If so, the spark module is probably OK. What typically goes wrong with the sparking system is that the rotary switch on the valve stops working. Test continuity as described in section 5-3(a). If that isn't the problem, check the electrode for damage and proper adjustment. The spark target (the nearest metal to the electrode) should be about 1/8" to 3/16" away from it, (about the thickness of 2-3 dimes) and directly across the primary pilot orifice. Replace or adjust the electrode as appropriate. When replacing, make sure you get the right kind of electrode (there are several) and do not cut the electrode lead; follow it all the way back to the spark module and plug the new lead into the proper spark module terminal.

IF SPARKING DOES NOT STOP

Usually the ignition switch has gotten some moisture in it and it is shorting. To test, pull the lead off the switch and see if the sparking stops. If so, the switch is bad. Replace it.

In certain ovens, a failure in the spark module will cause this symptom. If the switch does not test defective, replace the spark module.

Remember that these switches are on 110 volt circuits. If you get too fast and loose with pulling these leads off to test them, you might zap yourself.

Figure 6-H Pilot Valve Spark Oven Ignition System

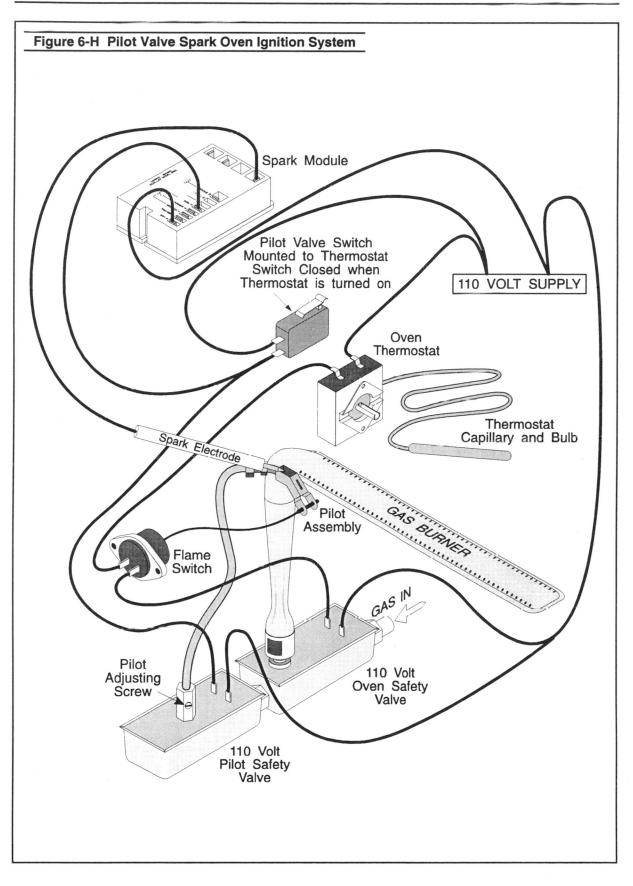

Spark Module

Pilot Valve Switch
Mounted to Thermostat
Switch Closed when
Thermostat is turned on

110 VOLT SUPPLY

Oven
Thermostat

Thermostat
Capillary and Bulb

Spark Electrode

GAS BURNER

Pilot
Assembly

GAS IN

Flame
Switch

Pilot
Adjusting
Screw

110 Volt
Oven Safety
Valve

110 Volt
Pilot Safety
Valve

6-4 GLOW-BAR IGNITION

A glow-bar ignitor is simply a 110 volt heating element that glows yellow-hot, well more than hot enough to ignite the gas when the gas touches it. It is wired *in series* with the oven safety valve.

When two electrical components are wired in series, they share the voltage (see figure 6-I) according to how much resistance they have. If the ignitor has, say, two-thirds of the resistance of the entire circuit, it will get two-thirds of the voltage. In reality, the ignitor has *most* of the resistance, so it gets *most* of the voltage. Out of 110 volts, the safety valve only gets about 3 to 4 volts. So unlike the flame safety switch system (which has a safety valve that looks almost exactly like this one,) this safety valve is a *low voltage* valve.

Let's talk about honey for a minute. Yes, honey. Stick with me here. (Pardon the pun)

If you take a squeeze bottle full of honey out of a refrigerator and try to squeeze some honey out of it, it is difficult. There is a lot of resistance, because the honey is cold and thick. If you heat up that honey in the microwave for a few seconds, it becomes thinner and flows easier; you can get more honey out of the bottle more quickly.

The same thing happens, electrically speaking, with the ignitor. When you first apply voltage, the ignitor is cold and the resistance is high. When the ignitor heats up, the resistance drops, and electricity is able to flow through it more easily, and the voltage across it drops.

Now apply that fact to what we just said about the ignitor and gas valve splitting the voltage. When the ignitor is cold, its resistance is high, and it gets most of the voltage. *So much, in fact, that there isn't enough voltage left to open the safety valve.* When the ignitor heats up, the resistance drops, the safety valve gets more voltage, and bingo! The

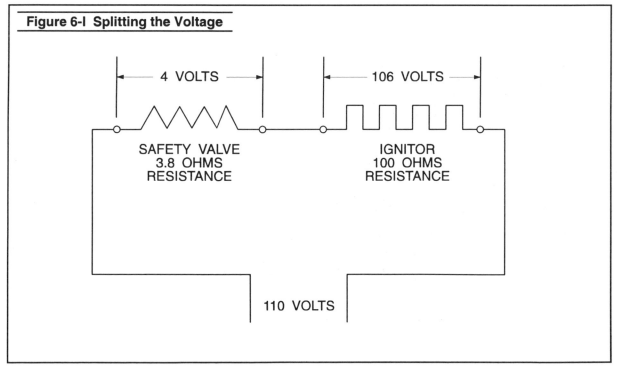

Figure 6-I Splitting the Voltage

← 4 VOLTS →   ← 106 VOLTS →

SAFETY VALVE
3.8 OHMS
RESISTANCE

IGNITOR
100 OHMS
RESISTANCE

110 VOLTS

safety valve opens, and the ignitor is hot enough to ignite the gas. *Note that the ignitor stays on the whole time the burner is burning.*

The safety valve contains a small heater and a bimetal connected directly to the gas valve. (see figure 6-J) When enough voltage is applied, the heater heats the bimetal, which opens the valve. When the oven thermostat senses the

right oven temperature, it cuts off the power to the ignitor and gas valve, and the bimetal cools and shuts off the gas flow.

There are two different types of ignitors in common use, round (aka carborundum) and flat (aka square, or Norton-type.) (see figure 6-K) Their resistances are different, so the gas safety valve that each uses is different, but the principle is the same.

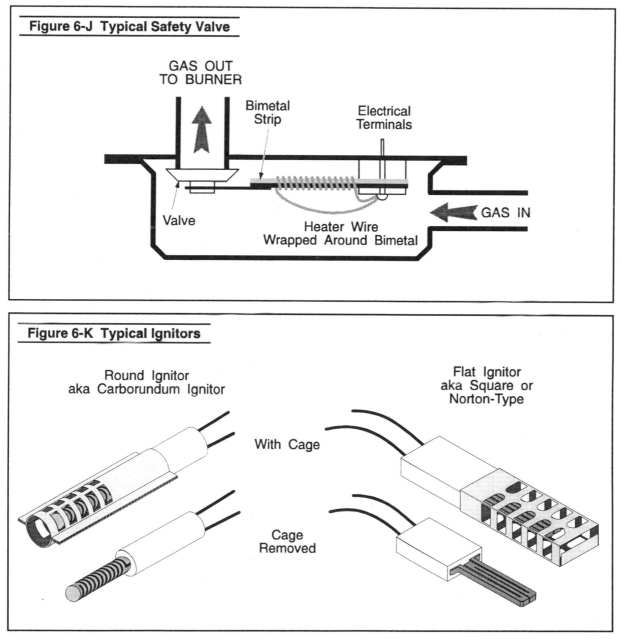

Figure 6-J Typical Safety Valve

Figure 6-K Typical Ignitors

TROUBLESHOOTING & REPAIR OF GLOW-BAR IGNITION SYSTEMS

The symptom, of course, is that the oven burner won't ignite. As always, first check the automatic bake cycle (timer) controls, if any, as mentioned in section 2-6(d).

Aside from the thermostat and other safety controls, there really are only two parts to the glow-bar ignition system; the ignitor and the safety valve. (Figure 6-L)

There may be a fuse as well. Usually the fuse is located down near the safety valve, but in some installations it's under the cooktop or inside the console. If you have a system with a fuse, check that first for continuity.

Typically it's the ignitor that fails, and when it does, you can usually see the damage to the element inside the cage.

However, this is not always true. If the ignitor gets old and weak, it will still glow, but usually only red- or orange-hot, not yellow-hot. When this happens, the resistance is still too high for the safety valve to open. It can throw you; since you see the ignitor working, you think the problem is the safety valve. It's usually not. And electrical parts are non-returnable.

When diagnosing these systems, I recommend replacing the ignitor first. If the problem turns out to be the gas valve, you just ate an ignitor, but trust me, you'd have needed one sooner or later anyway.

It is very important to make sure you get the right ignitor and/or gas valve for your oven. The safety valves are matched to the ignitor, and the ignitor is matched to the burner.

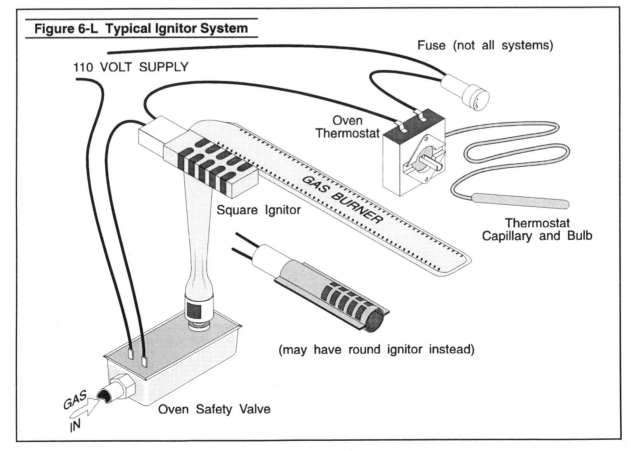

Figure 6-L Typical Ignitor System

Fuse (not all systems)

110 VOLT SUPPLY

Oven Thermostat

Square Ignitor

GAS BURNER

Thermostat Capillary and Bulb

(may have round ignitor instead)

GAS IN

Oven Safety Valve

6-5 PILOTS AND BURNERS MAINTENANCE & REPAIR

Natural gas, they tell us, is clean burning. Well, it makes good ad copy, but it's not 100 percent true. There are trace impurities in gas, and when they burn they become ash. And over a long long period of time this ash can build up and clog tiny gas orifices, like pilot orifices. The symptoms may be that the pilot will not stay lit, or blows out too easily. In an oven, it usually also means that the pilot will not get hot enough to open the safety valve, and the burner will not light. Or the ash might build up on the pilot sensing bulb, and insulate it enough that the safety valve operates intermittently or not at all.

You can usually clean them out with an old toothbrush and some compressed air, but pilot orifices are generally so inexpensive that it's cheaper and safer to just replace them. If you choose to clean them out, use a soft-bristle brush like a toothbrush, and not a wire brush; a wire brush might damage the orifice. Be careful not to push the ash into the orifice and impact it.

Oven burners are not like surface burners; rarely are they accidentally contacted by food. If they have any trouble at all, it's usually a little ash around the holes. Clean as described above. If you do have any crusty stuff from accidental contact with food, clean as described in section 5-4.

Chapter 7

DIGITAL OVEN CONTROL FAULT OR ERROR CODES

If your oven has digital controls and it is displaying a fault or error code, (F1, F2, E3, E8, etc) it usually boils down to one of two or three problem areas. Either the temperature sensor, or something in the control circuitry has gone bad. There may also be a problem with the door switch, or the door locking mechanism or switch in a self-cleaning oven.

Usually you will find the sensor inside the oven, at the top, near the back. The sensor may be removed through either the front or the back of the oven. Temperature sensors are usually fairly easy and inexpensive to identify and replace.

Not so with the solid state circuit components. You may be able to trace the problem to a defective keypad or EOC or ERC (Electronic Oven / Range Control; essentially the clock unit,) but often the diagnosis is ambiguous. (i.e., it's either the ERC or the keypad.)

Unfortunately, keypads and ERC's are electrical parts, and *expensive* ones at that. Electrical parts are almost always non-returnable at the parts store. So if you've misdiagnosed a problem and bought a keypad when the problem was a circuit board, you're now the not-so-proud owner of an expensive keypad that is of no use to you.

For this reason, I caution you: IF YOU ARE NOT SURE OF YOUR DIAGNOSIS, CALL A QUALIFIED, FACTORY-AUTHORIZED TECHNICIAN to diagnose and repair your oven. It may cost a little more now, but ultimately you will probably save yourself a LOT of hassle and money. The factory-authorized technician has an advantage in that he *can* diagnose by trial and error and reuse the part.

Please note also that manufacturers constantly change things, using different parts on different oven models. Though we do our level best to keep up on all the changes, we do not guarantee the accuracy of the error codes listed. Use them at your own risk.

And don't forget; you're dealing with electricity here. As always, disconnect power on any control you're disassembling or assembling. When you need to test a live circuit, always make sure all contacts are clear from shorting, then energize the circuit just long enough to do your testing, Then de-energize the circuit again. Safety first!

Sears / Kenmore Ranges and Ovens

In my never-to-be-humble opinion, Sears is a great company. They have a terrific marketing department; for most people, the name Kenmore evokes all kinds of emotion. However, I need to point out that Sears is a department store. They don't build ovens.

What they do is to take ovens that other companies build and put the Kenmore name on them. That's a fine strategy, but when the machine has a problem, it can make diagnosing and finding the correct replacement part something of a task.

Correctly diagnosing a fault code in a Sears / Kenmore oven starts with identifying who made the oven, using the first three digits of the model number. The following lists the manufacturers of these machines, though it should not be considered exhaustive:

103	Roper	484	Fedders	665	Whirlpool
106	Whirlpool	562	Whirlpool	747	Litton
110	Whirlpool	562	Toshiba	790	Westinghouse
155	Roper	587	D & M	791	Tappan
174	Caloric	596	Amana	835	Roper
198	Whirlpool	628	Kelvinator	911	Roper
253	Gibson	629	Jenn Aire	925	Maycor
362	GE	647	Roper	960	Caloric
363	GE	651	Speed Queen		
417	Kelvinator	662	Kelvinator		

Whirlpool, Kitchenaid, Roper Codes

4-Digit Display

Fault Code	Problem	Diagnosis / Repair
F0 -E0	Analog to digital control failure	Shut off the breaker to the oven for 30 seconds If the error code display reappears, Replace the control
F1-E1	Safety flip-flop	Replace the ERC
F2-E0	Shorted Keypad	Replace the Keypad

Whirlpool, Kitchenaid, Roper Codes (continued)

F3-E0
    Sensor or sensor fuse open   Replace the sensor or fuse

F3-E1, F3-E2, F3-E3
    The sensor is bad, either shorted,   Replace the sensor
    oven temp or cleaning temp is too hot

F5-E0, F5-E1, F5-E2
    Problem with the door circuit   Usually the door latch or switch is bad,
                                        also check wiring

3-Digit Display

Fault Code	**Problem**	**Diagnosis / Repair**
F0, F1, F5	Failed ERC component	Replace the ERC
F2	Oven temperature is too high	Replace Sensor
F3, F4	Sensor or sensor fuse open or shorted	Replace the sensor or fuse
F6	Problem with the time circuit	Reset cooking time, or check for proper ground
F7, F8	Failure or stuck button on ERC	Replace the ERC
F9	Problem with the door circuit	Usually the door latch or switch is bad, also check wiring

General Electric, Hotpoint, RCA Codes
also Frigidaire and Tappan

Fault Code	Problem	Diagnosis / Repair
F0 or F1	Failed ERC Component	Replace the ERC
F2	Oven too hot	Bad sensor
F3	Open sensor circuit	Bad sensor, sensor fuse or wiring
F4	Shorted sensor or wiring circuit	Replace the sensor or find the short
F5, F6, F7 (Frigidaire, Tappan)	Component failure in the ERC	Replace the ERC
F7 (GE, Hotpoint, RCA)	Either the clock is bad or there is a button stuck on the keypad	Call a factory-authorized technician
F8 (GE, Hotpoint, RCA)	Component failure in the ERC	Replace the ERC
F8 (Frigidaire, Tappan)	Problem with the door circuit	Check the wiring and door switch(es)
F9	Problem with the door circuit	Check the wiring and door switch(es)

General Electric, Hotpoint, RCA "New" Codes ('99)

Fault Code	Problem	Diagnosis / Repair
FFF		
	Control Error, Failed EEPROM	Replace the control
F0, F7		
	Either the ERC is bad or there is a button stuck on the keypad	Call a factory-authorized technician
F2		
	During Bake - Oven too hot	Bad sensor, or bake relay contacts
	During Clean - Oven too hot	Bad sensor, or clean relay contacts
		Both locking switches closed at the same time
F3		
	Open sensor circuit	Bad sensor, sensor fuse or wiring
F4		
	Shorted sensor or wiring circuit	Replace the sensor or find the short
F5		
	Loss of relay circuit	Call a factory-authorized technician
FC, FF		
	Door motor safety circuit	ERC

Maycor: Maytag, Magic Chef, Admiral, Jenn-Air

Far and away, the most common problem on these ranges is a flashing F1 error code. To diagnose, first remove power from the range, by unplugging it or tripping the breaker. Open the control panel and disconnect the touchpad ribbon connector from the ERC. Make sure nothing will short out, and put power back on the range.

If the range is beeping and still showing "F1," the ERC is bad; replace it.

If the time appears, or if the clock is blank, the touchpad is bad; replace it.

Fault Code	Problem	Diagnosis / Repair
F1	Bad ERC or touchpad	Test as described above, and replace the bad component
F2	Oven too hot	Bad relay board or bad sensor
F3, F4	Open or shorted sensor	Replace the sensor
F5	Hardware and watchdog circuits disagree	Replace the clock
F6	Missing AC Signal a button stuck on the keypad	Call a factory-authorized technician
F7	Function key shorted or a button is stuck	Try to unstick the button or replace the touchpad
F8	Problem with A/D converter	Replace the clock
F9	Door locking problem	Check the door locking circuit
F10	Function key stuck	Try to unstick the button or replace the touchpad

Caloric and Amana Codes

***for Glass Link ERC; also see ERCIII below**

<u>Fault Code</u>	<u>Problem</u>	<u>Diagnosis / Repair</u>
Door switch flashing a number 0, 1, 2 or 3, or lock flashing		
	Door latch switch broken	Replace
F0	No safety signal	Replace the adapter board
F1	Watchdog circuit	Replace the ERC
F2	Oven too hot	Bad sensor
F3,F4	Open or shorted sensor	Replace the sensor
F7	The touch panel is shorted	Replace the glass panel
F9	Motorized door	Check the door latch
FF	Door Locking Error	Check the wiring and door switch(es)

***for ERC III / Relay board with separate board and push buttons**

F0	Shorted push pad	Replace the push pad
F1, F5, F6, F8	Component failure on ERC	Replace the ERC
F2	Oven too hot	Bad sensor - check for 1100 ohms resistance
F3,F4	Open or shorted sensor	Replace the sensor
F7	A button on the touchpad is shorted or sticking	Check the glass alignment
F9	Door latch failure	Replace

Thermador

First, try to reset the error code by turning power off to the oven, waiting 30 seconds, then turning power back on.

Error
Code Notes   Problem

Code	Notes	Problem
E1	C,E	Control board problem
E2	G,K	Cook or clean mode runaway
E3	A,H	Open sensor
E4	A,H	Shorted sensor
E5	D,K	Control Board too hot or too cold, or defective board
E6	C,J	Control Board Problem
E7	A,I	Illegal temp display (turn off and retry)
E8	A,D	Control Board Problem
E9	A,E	Latch Switch Problems
E10	B,D	Control Board Problem
E11	A,F	CT Oven; Latch Switch Problem
E12	A,E	CT Oven; Latch Switch Problem
E13	A,C	CT Oven; Latch Switch Frozen or no power to latch motor
		CMT Ovens; control board not converted
E14	A,E	Latch Switch Problem
E15	B,D	Control Board Problem

Notes

A	Turns heat off on failed oven only; but the microwave is not affected
B	Turns all heat off; but the microwave not affected
C	Disables the clean mode in both ovens; cook mode and micro are not affected
D	Error will remain in display until repaired and powered up. No error tones.
E	Turning to "off" stops the error and flashing can tweak away the code.
F	Tweaking turns to "---" for retry.
G	Clears when oven temp drops below the runaway temp and the selector is off.
H	Can be cancelled by tweaking if a good sensor is detected.
I	Cleared with a mode change
J	Can be tweaked away for retry.
	User must unlatch and delete the "---" to try to relatch the door.
K	Turns all heat and microwave off.
L	If two switches show locked door, then "E13" and "lock" are permanent in the display (all modes). If two switches show an open door, you can tweak away the "E13."

Index